Andre Jean Marie Hamon

Meditations For All The Days Of The Year

Andre Jean Marie Hamon

Meditations For All The Days Of The Year

ISBN/EAN: 9783744653329

Printed in Europe, USA, Canada, Australia, Japan

Cover: Foto ©berggeist007 / pixelio.de

More available books at **www.hansebooks.com**

MEDITATIONS

FOR ALL THE DAYS OF THE YEAR.

FOR THE USE OF PRIESTS, RELIGIOUS,
AND THE FAITHFUL.

BY

REV. M. HAMON, S.S.,

Pastor of St. Sulpice, Paris, Author of "Life of St. Francis de Sales"
and "Life of Cardinal Cheverus."

From the Twenty-third Revised and Enlarged Edition

BY

MRS. ANNE R. BENNETT,
(*née* GLADSTONE.)

THIRD EDITION.
WITH A METHOD OF USING THESE MEDITATIONS.
BY VERY REV. A. MAGNIEN, S.S., D.D.

VOLUME II.

*From Septuagesima Sunday to the Second Sunday after
Easter.*

NEW YORK, CINCINNATI, CHICAGO:
BENZIGER BROTHERS.
...... to the Holy Apostolic See

Nihil Obstat.

   D. J. McMahon, D. D.,

        Censor Librorum.

Imprimatur.

   ✠ MICHAEL AUGUSTINE,

       Archbishop of New York

New York, July 14, 1894.

     Copyright, 1894, by Benziger Brothers.
    Printed in the United States of America

CONTENTS.

	PAGE
MORNING PRAYERS	9
EVENING PRAYERS	17

SEPTUAGESIMA SUNDAY.—GOD CALLS UPON US TO SERVE HIM.................... 23

MONDAY.—Obligation and Recompense in the Service of God.................... 30
TUESDAY.—Jesus in the Garden of Olives............. 34
WEDNESDAY.—All Creation Invites us to Serve God.... 38
THURSDAY.—How we must make use of Creatures for the Service of God.................... 43
FRIDAY.—The Duty, Honor, and Glory of the Service of God.................... 47
SATURDAY.—The Small Number of the Elect........... 51

SEXAGESIMA SUNDAY.—THE EXCELLENCE OF THE WORD OF GOD.................... 56

MONDAY.—Obstacles to the Word of God............. 61
TUESDAY.—The Passion of Our Lord................. 65
WEDNESDAY.—The Respect and Attention we owe to the Word of God 70

	PAGE
THURSDAY.—The Means of Profiting by the Word of God.	74
FRIDAY.—The Reading of the Scriptures	78
SATURDAY.—Spiritual Reading	83

QUINQUAGESIMA SUNDAY.—THE CARNIVAL SEASON........ 88

MONDAY.—The Ingratitude of Men to Our Lord	93
TUESDAY.—The Gentleness and Humility of Jesus Christ under Outrage	97
ASH WEDNESDAY.—The Thought of Death	101
THURSDAY.—The Ashes a Lesson of Humility	105
FRIDAY.—Jesus Crowned with Thorns	109
SATURDAY.—Holiness of the Season of Lent	113

FIRST SUNDAY IN LENT.—TEMPTATIONS IN GENERAL........ 117

MONDAY.—The Three Temptations of Jesus in the Desert	123
TUESDAY.—The Sacrament of Penance	127
WEDNESDAY.—The Examination of Conscience	132
THURSDAY.—The Manner of Making the Examination	135
FRIDAY.—The Devotion to the Nails and Lance of the Passion	139
SATURDAY.—The Particular Examination	143

SECOND SUNDAY IN LENT.—THE TRANSFIGURATION........ 147

MONDAY.—The Necessity and Method of Prayer	152
TUESDAY.—The Greatness of Jesus Revealed on Thabor.	157

	PAGE
WEDNESDAY.—Holy Desires after Heaven	161
THURSDAY.—The Love of Suffering	165
FRIDAY.—The Holy Winding-sheet	170
SATURDAY.—Lessons of Humility and of Detachment upon Thabor	174

THIRD SUNDAY IN LENT.—THE SIN OF RELAPSE............ 178

MONDAY.—Interior and Universal Contrition	183
TUESDAY.—Supreme and Supernatural Contrition	187
WEDNESDAY.—The Offence Committed against God by Sin	191
THURSDAY.—The Hatred God has for Sin	196
FRIDAY.—The Five Wounds	201
SATURDAY.—Mid-Lent	206

FOURTH SUNDAY IN LENT.—MULTIPLICATION OF THE LOAVES.... 210

MONDAY.—The Evil Sin Causes us	215
TUESDAY.—A Firm Resolve	219
WEDNESDAY.—Confession	223
THURSDAY.—Satisfaction	228
FRIDAY.—The Precious Blood	233
SATURDAY.—On Direction	237

PASSION SUNDAY.—HOW JESUS CRUCIFIED LOVES US........ 241

MONDAY.—How we ought to Love Jesus Crucified	247
TUESDAY.—The Cross the Salvation and Consolation of the Christian	251

WEDNESDAY.—The Cross the Strength and Glory of the Christian.. 256
THURSDAY.—The Cross the Science of the Christian ... 261
FRIDAY.—The Compassion of the Blessed Virgin....... 266
SATURDAY.—The Cross the Science of the Christian.... 271

PALM SUNDAY.—TRIUMPHANT ENTRANCE OF JESUS INTO JERUSALEM................ 275

MONDAY IN HOLY WEEK.—What Christ Suffered from His Apostles... 281
TUESDAY IN HOLY WEEK.—What Christ Suffered from His Enemies... 287
WEDNESDAY IN HOLY WEEK.—Jesus on Calvary....... 291
THURSDAY IN HOLY WEEK.—The Institution of the Eucharist and of the Priesthood.................... 295
GOOD FRIDAY.—Love and Conversion................. 300
SATURDAY IN HOLY WEEK.—The Burial of the Saviour and His Descent into Limbo...................... 304

EASTER DAY.—THE RESURRECTION THE TRIUMPH OF FAITH AND OF HOPE..... 308

EASTER MONDAY.—The Disciples at Emmaus.......... 314
EASTER TUESDAY.—The Apparition of Jesus to the Apostles... 321
WEDNESDAY.—His Apparition on the Borders of the Lake of Tiberias..................................... 326
THURSDAY.—His Apparition to Mary Magdalene....... 332
FRIDAY.—His Apparition upon a Mountain of Galilee... 337
SATURDAY.—His Apparition to the Holy Women and to the Apostles.. 343

	PAGE
LOW SUNDAY.—INTERIOR PEACE	347

MONDAY.—Excessive Activity an Obstacle to Peace..... 354
TUESDAY.—Other Obstacles, Preoccupation and Discouragement 358
WEDNESDAY.—Other Obstacles, Vain Joy and Wrong Sadness... 362
THURSDAY.—Other Obstacles, Temptations and Scruples. 366
FRIDAY.—Means whereby to Attain Peace: Humility and Renunciation of a Sensual Life................ 370
SATURDAY.—Another Means whereby to Attain Peace: Conformity to the Will of God................... 374

SAINTS

WHOSE FEASTS ARE CELEBRATED ON FIXED DAYS, AND WHICH THEREFORE DO NOT FOLLOW THE VARIABLE COURSE OF THE LITURGY.

FEBRUARY TWENTY-FOURTH.—St. Matthias............. 379
MARCH NINETEENTH.—St. Joseph.—His Glory......... 382
MARCH TWENTIETH.—Confidence in St. Joseph........ 387
MARCH TWENTY-FIRST.—Graces Attached to the Devotion to St. Joseph........ 391
MARCH TWENTY-SECOND.—Hidden Life of St. Joseph... 397
MARCH TWENTY-THIRD.—Tried Life of St. Joseph...... 401
MARCH TWENTY-FIFTH.—The Annunciation........... 407

Morning Prayers.

In the name of the Father, ✠ and of the Son, and of the Holy Ghost. Amen.

Place Yourself in the Presence of God, and adore His holy Name.

Most holy and adorable Trinity, one God in three Persons, I believe that Thou art here present: I adore Thee with the deepest humility, and render to Thee, with my whole heart, the homage which is due to Thy sovereign majesty.

An Act of Faith.

O my God, I firmly believe that Thou art one God in three divine Persons, Father, Son, and Holy Ghost; I believe that Thy divine Son became man, and died for our sins, and that He will come to judge the living and the dead. I believe these and all the truths which the holy Catholic Church teaches, because Thou hast revealed them, who canst neither deceive nor be deceived.

An Act of Hope.

O my God, relying on Thy infinite goodness and promises, I hope to obtain pardon of my sins, the help of Thy grace, and life everlasting, through the merits of Jesus Christ, my Lord and Redeemer.

An Act of Love.

O my God, I love Thee above all things, with my whole heart and soul, because Thou art all-good and worthy of all love. I love my neighbor as myself for the love of Thee. I forgive all who have injured me, and ask pardon of all whom I have injured.

Thank God for All Favors and Offer Yourself to Him.

O my God, I most humbly thank Thee for all the favors Thou hast bestowed upon me up to the present moment. I give Thee thanks from the bottom of my heart that Thou hast created me after Thine own image and likeness, that Thou hast redeemed me by the precious blood of Thy dear Son, and that Thou hast preserved me and brought me safe to the beginning of another day. I offer to Thee, O Lord, my whole being, and in particular all my thoughts, words, actions, and sufferings of this day. I consecrate them all to the glory of Thy name, beseeching Thee that through the infinite merits of Jesus Christ my Saviour they may all find acceptance in Thy sight. May Thy divine love animate them, and may they all tend to Thy greater glory.

Resolve to Avoid Sin and to Practise Virtue.

Adorable Jesus, my Saviour and Master, model of all perfection, I resolve and will endeavor this day to imitate Thy example, to be, like Thee, mild, humble, chaste, zealous, charitable, and resigned. I will redouble my efforts that I may not fall this day into any of those sins which I have heretofore committed (*here name any besetting sin*), and which I sincerely desire to forsake.

Ask God for the Necessary Graces.

O my God, Thou knowest my poverty and weakness, and that I am unable to do anything good without Thee; deny me not, O God, the help of Thy grace; proportion it to my necessities; give me strength to avoid anything evil which Thou forbiddest, and to practise the good which Thou hast commanded; and enable me to bear patiently all the trials which it may please Thee to send me.

The Lord's Prayer.

Pater noster, qui es in cœlis, sanctificetur nomen tuum: adveniat regnum tuum ; fiat voluntas tua, sicut in cœlo, et in terra. Panem nostrum quotidianum da nobis hodie: et dimitte nobis debita nostra, sicut et nos dimittimus debitoribus nostris. Et ne nos inducas in tentationem ; sed libera nos a malo. Amen.

Our Father, who art in heaven, hallowed be Thy name: Thy kingdom come: Thy will be done on earth, as it is in heaven. Give us this day our daily bread: and forgive us our trespasses, as we forgive those who trespass against us. And lead us not into temptation; but deliver us from evil. Amen.

The Hail Mary.

Ave, Maria, gratia plena: Dominus tecum: benedicta tu in mulieribus, et benedictus fructus ventris tui, Jesus. Sancta Maria, Mater Dei, ora pro nobis peccatoribus, nunc et in hora mortis nostræ. Amen.

Hail, Mary full of grace; the Lord is with thee; blessed art thou among women, and blessed is the fruit of Thy womb, Jesus. Holy Mary, Mother of God, pray for us sinners, now and at the hour of our death. Amen.

The Apostles' Creed.

Credo in Deum, Patrem omnipotentem, Creatorem cœli et terræ; et in Jesum Christum, Filium ejus unicum, Dominum nostrum; qui conceptus est de Spiritu Sancto, natus ex Maria Virgine, passus sub Pontio Pilato, crucifixus, mortuus et sepultus. Descendit ad inferos; tertia die resurrexit a mor-

I believe in God, the Father Almighty, Creator of heaven and earth; and in Jesus Christ, His only Son, our Lord: who was conceived by the Holy Ghost, born of the Virgin Mary, suffered under Pontius Pilate, was crucified, died, and was buried. He descended into hell; the third

tuis; ascendit ad cœlos, sedet ad dexteram Dei Patris omnipotentis; inde venturus est judicare vivos et mortuos. Credo in Spiritum Sanctum, sanctam Ecclesiam Catholicam, sanctorum communionem, remissionem peccatorum, carnis resurrectionem, vitam æternam. Amen.

day He rose again from the dead; He ascended into heaven, and sitteth at the right hand of God, the Father Almighty; from thence He shall come to judge the living and the dead. I believe in the Holy Ghost, the holy Catholic Church, the communion of saints, the forgiveness of sins, the resurrection of the body, and the life everlasting. Amen.

Ask the Prayers of the Blessed Virgin, your Guardian Angel, and your Patron Saint.

Holy Virgin, Mother of God, my Mother and Patroness, I place myself under thy protection, I throw myself with confidence into the arms of thy compassion. Be to me, O Mother of mercy, my refuge in distress, my consolation under suffering, my advocate with thy adorable Son, now and at the hour of my death.

> Angel of God, my guardian dear,
> To whom His love commits me here,
> Ever this day be at my side,
> To light and guard, to rule and guide. Amen.

O great Saint whose name I bear, protect me, pray for me, that like thee I may serve God faithfully on earth, and glorify Him eternally with thee in heaven. Amen.

Litany of the Most Holy Name of Jesus.

Kyrie eleison.	Lord, have mercy on us.
Christe eleison.	Christ, have mercy on us.
Kyrie eleison.	Lord, have mercy on us.
Jesu audi nos.	Jesus, hear us.

Morning Prayers. 13

Jesu exaudi nos.	Jesus, graciously hear us.
Pater de cœlis Deus,	God the Father of heaven,
Miserere nobis.	*Have mercy on us.*
Fili, Redemptor mundi, Deus,	God the Son, Redeemer of the world, *Have, etc.*
Miserere nobis.	
Spiritus Sancte Deus,	God the Holy Ghost,
Sancta Trinitas, unus Deus,	Holy Trinity, one God,
Jesu, Fili Dei vivi,	Jesus, Son of the living God,
Jesu, splendor Patris,	Jesus, splendor of the Father,
Jesu, candor lucis æternæ,	Jesus, brightness of eternal light,
Jesu, rex gloriæ,	Jesus, king of glory,
Jesu, sol justitiæ,	Jesus, sun of justice,
Jesu, fili Mariæ Virginis,	Jesus, son of the Virgin Mary,
Jesu amabilis,	Jesus, most amiable,
Jesu admirabilis,	Jesus, most admirable,
Jesu, Deus fortis,	Jesus, mighty God,
Jesu, pater futuri sæculi,	Jesus, father of the world to come,
Jesu, magni consilii angele,	Jesus, angel of the great council,
Jesu potentissime,	Jesus, most powerful,
Jesu patientissime,	Jesus, most patient,
Jesu obedientissime,	Jesus, most obedient,
Jesu, mitis et humilis corde,	Jesus, meek and humble of heart,
Jesu, amator castitatis,	Jesus, lover of chastity,
Jesu, amator noster,	Jesus, lover of us,
Jesu, Deus pacis,	Jesus, God of peace,
Jesu, auctor vitæ,	Jesus, author of life,
Jesu, exemplar virtutum,	Jesus, model of virtues,
Jesu, zelator animarum,	Jesus, zealous for souls,
Jesu, Deus noster,	Jesus, our God,

Miserere nobis. — *Have mercy on us.*

Jesu, refugium nostrum,	Jesus, our refuge,
Jesu, pater pauperum,	Jesus, father of the poor,
Jesu, thesaurus fidelium,	Jesus, treasure of the faithful,
Jesu, bone pastor,	Jesus, good shepherd,
Jesu, lux vera,	Jesus, true light,
Jesu, sapientia æterna,	Jesus, eternal wisdom,
Jesu, bonitas infinita,	Jesus, infinite goodness,
Jesu, via et vita nostra,	Jesus, our way and our life,
Jesu, gaudium angelorum,	Jesus, joy of angels,
Jesu, rex patriarcharum,	Jesus, king of patriarchs,
Jesu, magister apostolorum,	Jesus, master of apostles,
Jesu, doctor evangelistarum,	Jesus, teacher of evangelists,
Jesu, fortitudo martyrum,	Jesus, strength of martyrs,
Jesu, lumen confessorum,	Jesus, light of confessors,
Jesu, puritas virginum,	Jesus, purity of virgins,
Jesu, corona sanctorum omnium,	Jesus, crown of all saints,

Miserere nobis. / *Have mercy on us.*

Propitius esto,	Be merciful,
Parce nobis, Jesu.	*Spare us, O Jesus.*
Propitius esto,	Be merciful,
Exaudi nos, Jesu.	*Graciously hear us, O Jesu.*
Ab omni malo,	From all evil,
Ab omni peccato,	From all sin,
Ab ira tua,	From Thy wrath,
Ab insidiis diaboli,	From the snares of the devil,
A spiritu fornicationis,	From the spirit of fornication,
A morte perpetua,	From everlasting death,
A neglectu inspirationum tuarum,	From the neglect of Thy inspirations,
Per mysterium sanctæ incarnationis tuæ,	Through the mystery of Thy holy incarnation,

Libera nos, Jesu. / *Jesus, deliver us.*

Morning Prayers.

Per nativitatem tuam,	Through Thy nativity,
Per infantiam tuam,	Through Thine infancy,
Per divinissimam vitam tuam,	Through Thy most divine life,
Per labores tuos,	Through Thy labors,
Per agoniam et passionem tuam,	Through Thine agony and passion,
Per crucem et derelictionem tuam,	Through Thy cross and dereliction,
Per languores tuos,	Through Thy faintness and weariness,
Per mortem et sepulturam tuam,	Through Thy death and burial,
Per resurrectionem tuam,	Through Thy resurrection,
Per ascensionem tuam,	Through Thine ascension,
Per gaudia tua,	Through Thy joys,
Per gloriam tuam,	**Through Thy glory,**
Agnus Dei, qui tollis peccata mundi,	Lamb of God, who takest away the sins of the world
Parce nobis, Jesu.	*Spare us, O Jesus.*
Agnus Dei, qui tollis peccata mundi,	Lamb of God, who takest away the sins of the world,
Exaudi nos, Jesu.	*Graciously hear us, O Jesus.*
Agnus Dei, qui tollis peccata mundi,	Lamb of God, who takest away the sins of the world,
Miserere nobis, Jesu.	*Have mercy on us, O Jesus.*
Jesu audi nos.	Jesus, hear us.
Jesu exaudi nos.	*Jesus, graciously hear us.*

(marginal: *Libera nos, Jesu.* / *Jesus, deliver us.*)

Oremus. — *Let us pray.*

Domine Jesu Christe, qui dixisti: Petite, et accipietis; quærite, et invenietis; pulsate, et aperietur vobis, quæsumus; da nobis peten-

O Lord Jesus Christ, who hast said: Ask, and ye shall receive; seek, and ye shall find; knock, and it shall be opened unto you; grant, we

tibus divinissimi tui amoris affectum, ut te toto corde, ore et opere diligamus, et a tua nunquam laude cessemus.

Sancti Nominis tui, Domine, timorem pariter et amorem fac nos habere perpetuum, quia nunquam tua gubernatione destituis quos in soliditate tuæ dilectionis instituis. Qui vivis et regnas, etc. Amen.

beseech Thee, unto us who ask, the gift of Thy most divine love, that we may ever love Thee with all our hearts, and in all our words and actions, and never cease from showing forth Thy praise.

Make us, O Lord, to have a perpetual fear and love of Thy holy Name; for Thou never failest to govern those whom Thou dost solidly establish in Thy love. Who livest and reignest, etc. Amen.

The Angelus Domini.

Angelus Domini nuntiavit Mariæ, et concepit de Spiritu Sancto.

Ave Maria, etc.

Ecce ancilla Domini: fiat mihi secundum verbum tuum.

Ave Maria, etc.

Et verbum caro factum est, et habitavit in nobis.

Ave Maria, etc.

The angel of the Lord declared unto Mary, and she conceived of the Holy Ghost.

Hail Mary, etc.

Behold the handmaid of the Lord: be it done unto me according to thy word.

Hail Mary, etc.

And the Word was made flesh, and dwelt among us.

Hail Mary, etc.

Oremus.

Gratiam tuam, quæsumus, Domine, mentibus nostris infunde: ut qui, angelo nun-

Let us pray.

Pour forth, we beseech Thee, O Lord! Thy grace into our hearts, that we, unto

tiante, Christi Filii tui incarnationem cognovimus, per passionem ejus et crucem ad resurrectionis gloriam perducamur. Per eumdem Christum Dominum nostrum. Amen.

whom the incarnation of Christ, Thy Son, was made known by the message of an angel, may, by His passion and cross, be brought to the glory of the resurrection. Through the same Christ our Lord. Amen.

Evening Prayers.

In the name of the Father, ✠ and of the Son, and of the Holy Ghost. Amen.

Come, O Holy Ghost, fill the hearts of Thy faithful, and kindle in them the fire of Thy love.

Place Yourself in the Presence of God and Humbly Adore Him.

O my God, I present myself before Thee at the end of another day, to offer Thee anew the homage of my heart. I humbly adore Thee, my Creator, my Redeemer, and my Judge! I believe in Thee, because Thou art Truth itself; I hope in Thee, because Thou art faithful to Thy promises; I love Thee with my whole heart, because Thou art infinitely worthy of being loved; and for Thy sake I love my neighbor as myself.

Return Thanks to God for All His Mercies.

Enable me, O my God, to return Thee thanks as I ought for all Thine inestimable blessings and favors. Thou hast thought of me and loved me from all eternity; Thou hast formed me out of nothing; Thou hast delivered up Thy beloved Son to the ignominious death of the cross for my redemption; Thou hast made me a member of Thy holy Church; Thou hast preserved me from falling into the abyss of eternal misery, when my sins had provoked Thee to punish me; Thou hast graciously continued to spare me, even though I have not ceased to offend Thee. What return, O my God, can I make for Thy innumerable blessings, and particularly for the favors of this day? O all ye saints and angels, unite with me in praising the God of mercies, who is so bountiful to so unworthy a creature.

Our Father. Hail Mary. I believe.

Ask of God Light to Discover the Sins Committed this Day.

O my God, sovereign judge of men, who desirest not the death of a sinner, but that he should be converted and saved, enlighten my mind, that I may know the sins which I have this day committed in thought, word, or deed, and give me the grace of true contrition.

Here Examine your Conscience; then Say:

O my God, I heartily repent and am grieved that I have offended Thee, because Thou art infinitely good and sin is infinitely displeasing to Thee. I humbly ask of Thee mercy and pardon, through the infinite merits of Jesus Christ. I resolve, by the assistance of Thy grace, to do penance for my sins, and I will endeavor never more to offend Thee.

The Confiteor.

Confiteor Deo omnipotenti, beatæ Mariæ semper Virgini, beato Michaeli Archangelo, beato Joanni Baptistæ, sanctis apostolis Petro et Paulo, omnibus sanctis, et tibi Pater, quia peccavi nimis cogitatione, verbo, et opere, mea culpa, mea culpa, mea maxima culpa. Ideo precor beatam Mariam semper Virginem, bea-	I confess to Almighty God, to blessed Mary, ever Virgin, to blessed Michael the Archangel, to blessed John the Baptist, to the holy apostles Peter and Paul, and to all the saints, and to you, Father, that I have sinned exceedingly in thought, word, and deed, through my fault, through my fault, through my

tum Michaelem Archangelum, beatum Joannem Baptistam, sanctos apostolos Petrum et Paulum, omnes sanctos, et te, Pater, orare pro me ad Dominum Deum nostrum.

Misereatur nostri Omnipotens Deus, et dimissis peccatis nobis, perducat nos ad vitam æternam. Amen.

Indulgentiam, ✠ absolutionem, et remissionem peccatorum nostrorum, tribuat nobis omnipotens et misericors Dominus. Amen.

most grievous fault. Therefore I beseech blessed Mary ever Virgin, blessed Michael the Archangel, blessed John the Baptist, the holy apostles Peter and Paul, all the saints, and you, Father, to pray to the Lord our God for me.

May Almighty God have mercy upon us, and forgive us our sins, and bring us unto life everlasting. Amen.

May the Almighty and merciful Lord grant us pardon, ✠ absolution, and remission of our sins. Amen.

Pray for the Church of Christ.

O God, hear my prayers on behalf of our Holy Father Pope *N.*, our Bishops, our clergy, and for all that are in authority over us. Bless, I beseech Thee, the whole Catholic Church, and convert all heretics and unbelievers.

Pray for the Living and for the Faithful Departed.

Pour down Thy blessings, O Lord, upon all my friends, relations, and acquaintances, and upon my enemies, if I have any. Help the poor and sick, and those who are in their last agony. O God of mercy and goodness, have compassion on the souls of the faithful in purgatory; put an end to their sufferings, and grant to them eternal light, rest, and happiness. Amen.

Commend Yourself to God, to the Blessed Virgin, and the Saints.

Bless, O Lord, the repose I am about to take, that, my bodily strength being renewed, I may be the better enabled to serve Thee.

Evening Prayers.

O blessed Virgin Mary, Mother of mercy, pray for me that I may be preserved this night from all evil, whether of body or soul. Blessed St. Joseph, and all ye saints and angels of Paradise, especially my guardian angel and my chosen patron, watch over me. I commend myself to your protection now and always. Amen.

Litany of the Blessed Virgin.

Kyrie eleison.	Lord, have mercy.
Christe eleison.	Christ, have mercy.
Kyrie eleison.	Lord, have mercy.
Christe audi nos.	Christ, hear us.
Christe exaudi nos.	Christ, graciously hear us.
Pater de cœlis Deus, *miserere nobis*.	God the Father of heaven, *have mercy on us*.
Fili Redemptor mundi Deus, *miserere nobis*.	God the Son, Redeemer of the world, *have mercy on us*.
Spiritus Sancte Deus, *miserere nobis*.	God the Holy Ghost, *have mercy on us*.
Sancta Trinitas, unus Deus, *miserere nobis*.	Holy Trinity, one God, *have mercy on us*.
Sancta Maria, *ora pro nobis*.	Holy Mary, *pray for us*.
Sancta Dei genitrix,	Holy Mother of God,
Sancta virgo virginum,	Holy virgin of virgins,
Mater Christi,	Mother of Christ,
Mater divinæ gratiæ,	Mother of divine grace,
Mater purissima,	Mother most pure,
Mater castissima,	Mother most chaste,
Mater inviolata,	Mother inviolate,
Mater intemerata,	Mother undefiled,
Mater amabilis,	Mother most amiable,
Mater admirabilis,	Mother most admirable,
Mater Creatoris,	Mother of our Creator,
Mater Salvatoris,	Mother of our Saviour,
Virgo prudentissima,	Virgin most prudent,

Ora pro nobis. — *Pray for us.*

Evening Prayers.

Virgo veneranda,		Virgin most venerable,	
Virgo prædicanda,		Virgin most renowned,	
Virgo potens,		Virgin most powerful,	
Virgo clemens,		Virgin most merciful,	
Virgo fidelis,		Virgin most faithful,	
Speculum justitiæ,		Mirror of justice,	
Sedes sapientiæ,		Seat of wisdom,	
Causa nostræ lætitiæ,	*Ora pro nobis.*	Cause of our joy,	*Pray for us.*
Vas spirituale,		Spiritual vessel,	
Vas honorabile,		Vessel of honor,	
Vas insigne devotionis,		Singular vessel of devotion,	
Rosa mystica,		Mystical rose,	
Turris Davidica,		Tower of David,	
Turris eburnea,		Tower of ivory,	
Domus aurea,		House of gold,	
Fœderis arca,		Ark of the covenant,	
Janua cœli,		Gate of heaven,	
Stella matutina,		Morning star,	
Salus infirmorum,		Health of the sick,	
Refugium peccatorum,		Refuge of sinners,	
Consolatrix afflictorum,		Comforter of the afflicted,	
Auxilium Christianorum,		Help of Christians,	
Regina angelorum,		Queen of angels,	
Regina patriarcharum,	*Ora pro nobis.*	Queen of patriarchs,	*Pray for us.*
Regina prophetarum,		Queen of prophets,	
Regina apostolorum,		Queen of apostles,	
Regina martyrum,		Queen of martyrs,	
Regina confessorum,		Queen of confessors,	
Regina virginum,		Queen of virgins,	
Regina sanctorum omnium,		Queen of all saints,	
Regina sine labe originali concepta,		Queen conceived without original sin,	
Regina sacratissimi rosarii,		Queen of the most holy rosary,	

Agnus Dei, qui tollis peccata mundi, *parce nobis, Domine.*

Agnus Dei, qui tollis peccata mundi, *exaudi nos, Domine.*

Agnus Dei, qui tollis peccata mundi, *miserere nobis.*

V. Ora pro nobis, sancta Dei Genitrix.
R. Ut digni efficiamur promissionibus Christi.

Oremus.

Gratiam tuam, quæsumus, Domine, mentibus nostris infunde: ut qui, angelo nuntiante, Christi Filii tui incarnationem cognovimus, per passionem ejus et crucem ad resurrectionis gloriam perducamur. Per eumdem Christum Dominum nostrum. Amen.

Lamb of God, who takest away the sins of the world, *spare us, O Lord.*

Lamb of God, who takest away the sins of the world, *graciously hear us, O Lord.*

Lamb of God, who takest away the sins of the world, *have mercy on us.*

V. Pray for us, O holy Mother of God.
R. That we may be made worthy of the promises of Christ.

Let us pray.

Pour forth, we beseech Thee, O Lord, Thy grace into our hearts; that we, to whom the incarnation of Christ, Thy Son, was made known by the message of an angel, may, by His passion and cross, be brought to the glory of His resurrection. Through the same Christ our Lord. Amen.

MEDITATIONS

FOR ALL THE DAYS OF THE YEAR.

Septuagesima Sunday.

The Gospel according to St. Matthew, xx. 1-16.

"Jesus spoke this parable to His disciples: The kingdom of heaven is like to an householder, who went out early in the morning to hire laborers into his vineyard. And having agreed with the laborers for a penny a day, he sent them into his vineyard. And going out about the third hour, he saw others standing in the market-place idle. And he said to them: Go you also into my vineyard, and I will give you what shall be just. And they went their way. And again he went out about the sixth and the ninth hour, and did in like manner. But about the eleventh hour he went out and found others standing, and he saith to them: Why stand you here all the day idle? They say to him: Because no man hath hired us. He saith to them: Go you also into my vineyard. And when evening was come, the lord of the vineyard saith to his steward: Call the laborers and pay them their hire, beginning from the last even to the first. When therefore they were come that came about the eleventh hour they received every man a penny. But when the first also came, they thought that they should receive more; and they also received every man a penny. And receiving it they murmured against the master of the house, saying: These last have worked but one hour, and thou hast made them equal to us, that have borne the burden of the day and the heats. But he answering said to one of them: Friend, I do thee no wrong; didst thou not agree with me for a penny? Take what is thine, and go thy way: I will also give to this last even as to thee. Or, is it not lawful for me to do what I will? Is thy eye evil because I am good? So shall the last be first, and the first last. For many are called, but few chosen."

Summary of the Morrow's Meditation.

We will consecrate this week to meditations upon the gospel of Septuagesima Sunday, and to-morrow we will study the first words of it: "*Go you also into my vineyard*" (Matt. xx. 4). We shall learn thereby: 1st, that God obliges us to serve Him; 2d, how God wills that we should serve Him. We will then make the resolution: 1st, to employ every moment in doing what our conscience may prompt us to do in order to please God ; 2d, often to examine ourselves and ask whether what we do in regard to such or such a thing, our reading, our repasts, our visits, are indeed done for God and for love of Him. Our spiritual nosegay shall be the words of the Apostle : *" Whether you eat or drink, or whatsoever else you do, do all to the glory of God"* (I. Cor. x. 31).

Meditation for the Morning.

Let us adore God imposing upon all the precept of serving Him : go to my vineyard. Let us receive this precept with submission and love ; let us offer ourselves to God, to be forever His devoted servants ; and let us render Him homage as to our master.

FIRST POINT.

God obliges us to Serve Him.

To serve God is to employ our existence in doing what is pleasing to Him, and this obligation results from our belonging to Him wholly. He alone created us, He formed our members, He united them together so as to form of them a body ; He alone it was who animated this body by uniting with it a soul, endowed with the faculties of knowing, of willing, and of loving. He alone, consequently, is our master ; we are His property, His creature, His work, and we do not belong to ourselves. Now if the basis of our being belongs to God, all our acts ought equally to belong to Him, for the double reason that the revenues of a capital belong to the proprietor of the capital, and that God, in creating us, could not create us for any other end than that of serving Him, because there is no other end worthy of Him (Prov. xvi. 4). Therefore to seek ourselves or to seek the creature in whatever way we may is to commit a theft upon the essential domain of God. Therefore we ought not to live, to act, to speak, to think, except for God; not to use our feet except in order to go where He wills, or our hands except to do what He wills; our eyes, except to look at what He wills; our mind, except to think what He wills; our heart, except to love what He wills; our health, our strength, our time, except to employ them in what He wills; for all these things are His, and ought to serve only for what He wills. Therefore, whether I am in one condition or another, in suffering or enjoyment, in riches or in poverty, I have not the right to say anything against it. God is the master (I. Kings iii. 18). He can do with what belongs to Him as He pleases, and I ought always to find what He wills to be good. Oh, how does this truth confound me! for, alas! I think more of myself than I do of God, I work more for myself than for God, I love myself more than I love God. I forget that He is my end, that I ought to live only for Him; and as though I were myself my own end, I refer everything to myself, my comfort, my taste, my will. In thus turning myself away from my end, I compromise my salvation, my eternity. It is incumbent on me instantly to change my manner of life.

SECOND POINT.

How God wills that we should Serve Him.

God wills that we should give up ourselves to Him *wholly*, to Him *alone*, to Him *always*, to Him *by esteem and love.* 1st. To Him *wholly;* for since we have everything from Him, our body and our soul, and our faculties with all their acts, and our existence, together with every moment of which it is composed, we ought to give Him all; and in giving Him all we only give Him what belongs to Him already; to give Him the very least thing less than that would not satisfy Him (St. Prosper). 2d. To Him *alone;* for no one else having contributed to our being, except as the instrument of His will, I ought to serve Him alone, that is to say, with a constant and invariable intention, upright and pure, of pleasing Him alone, without having respect to any one else or to myself. To give to another the least portion of my heart or of my time would be the crime of the servant who, having in his hands the property of his master and the administration of his revenues, were to retain a portion of it for his own use or for that of his friends; for the acts of my body or my soul are as the products or the revenues of my substance which belongs entirely to God. 3d. To Him *always;* for all my moments belong to Him essentially; if He were to cease for a single moment to sustain me, I should fall into the abyss of nothingness; if He were to cease to concert with me as regards action, speech, or thought, I could not move, speak, or act. Therefore, every day and at every moment of the day and of the night I ought to be Thine, O my God, always endeavoring to please Thee; and to steal a single moment for myself, or for the creature, would be to injure Thy rights; it would be to usurp what belongs to Thee. 4th. I ought to give myself to God *by esteem and by love;* that is to say, that even if I were expecting nothing from God, I ought still to be wholly His, because He has created me and He preserves me through love which is wholly gratuitous, not only without interest, but often even against the interests of His glory which I offend. I ought therefore to forget myself in order to seek God alone in everything, and to do nothing except from love to Him. It is the first lesson of the catechism, contained in these words : God has created us in order to know Him, to love Him, and to serve Him ; such is the firm rock on which ought to be raised the edifice of all religion and of all perfection ; and it was with these thoughts present to him that Abraham found courage to quit his country, to sacrifice Isaac, and to lead a perfect life, and that Job found patience and resignation in the midst of the greatest calamities. It is for us to derive the same profit from them. Woe to us if we do not ! Yes, my God, I take my stand ; I am determined frankly, generously, entirely, to serve Thee ; I desire nothing but that in the whole world, and I desire it with my whole heart, free from any interested views and from any human respect. I leave Thee my heart, and I yield it up entirely to Thy love ; I devote it to Thy designs ; I abandon it to Thy guidance ; I will carefully avoid the least faults, and I will do all the good possible with all the perfection of which I am capable ; that is to say, promptly and without delay, fully and without any mixture in it of my own will, purely and without any other object than that of pleasing Thee, constantly and without becoming tired or wearied, or ceasing until I have finished what Thou willest of me.

Resolutions and spiritual nosegay as above.

Monday after Septuagesima.

Summary of the Morrow's Meditation.

We will continue to-morrow the same subject of meditation as that of this morning, and we shall see: 1st, that the God who calls us to serve Him has a right to require everything of us without promising us anything in return; 2d, that nevertheless He magnificently recompenses those who give Him all. We will then make the resolution: 1st, not to exercise any reserve in the service of God, and to grant to grace all that it asks of us; 2d, often to repeat to God in the form of an ejaculatory prayer that we are wholly His, and that we desire to live only for Him. Our spiritual nosegay shall be the first commandment of the Decalogue, *"Thou shalt love the Lord thy God with thy whole heart, with thy whole soul, and with thy whole strength"* (Deut. vi. 5).

Meditation for the Morning.

Let us adore God, our first principle and our last end, claiming through this double title all our services. As being made by Him we owe Him everything, as holding everything from Him we owe it all to Him a second time. Let us render

FIRST POINT.

God has a Right to Require All from us, without Promising us Anything in Return.

It being a fact that we derive everything from God, therefore in giving Him all we only render to Him that which belongs to Him. If amongst men benefits received are a sufficient reason for devotion to him by whom they have been given, even when no remuneration has been expected; if a man who is great in this world expects all from those whom he calls his creatures; if a father has a right to be loved and served by his children, even though he may have no inheritance and no money to leave them, and if such as these could not fail in their duty without attracting general reprobation and being looked upon as monsters of ingratitude, how much more ought we to be wholly belonging to God, without any thought of recompense! God has a right to say to us; "If you serve Me you will only have done your duty; I do not owe you anything on that account, no more than a father considers himself obliged to take into consideration the good offices done him by his son; but, on the other hand, if you do not serve Me as you ought and as My benefits oblige you to do, I will damn you." Legislators do not say : He who keeps the law shall be recompensed; they say : He who does not keep it shall be punished. The master does not say to his slave : Obey, and I will recompense you; he simply says : Obey, and no harm shall happen to you; if you do not obey, I will chastise you. Even admitting that all labor merits wages, God need not have promised more than a recompense as temporary as are our services; under no title does He owe us an eternal recompense, and if He promises it to us, it is pure goodness on His part; we should therefore be inexcusable if we were not wholly His, wholly His in all things, down to the smallest details of our conduct, always His through esteem and love.

SECOND POINT.

God Recompenses Magnificently those who give Him All.

God will give Himself to us in the same proportion that we give ourselves to Him. If He wills that we should be wholly His in a complete detachment from creatures, He also promises to be entirely ours. If He wills that we should be always His, He also wills to be always ours, as perfectly ours as though we were alone in the world. Our good God thirsts for our happiness. "Serve Me," He says; "think only of serving Me, and I will think of you, I will take care of you, I will give Myself to you as your possession and your treasure" (Gen. xv. 1). This doubtless has regard principally to eternity, but even in this present life what does He not do for those who give themselves fully and constantly to Him? He establishes His dwelling in their heart. He sheds therein His grace and His consolations. There is within such a heart a peace which surpasses all understanding, and which is accompanied with delicious joy (Gal. v. 22); it is like a continual feast (Prov. xv. 15); it is a foretaste of Paradise. Oh, how happy then are we when we love and serve God with all our heart! How miserable, on the contrary, when we resist the advances of God and make reservations in His service! We suffer greatly, we suffer without merit; and this anticipated hell is but the prelude of the other which will last eternally. O my God, how good it is then to serve Thee! (Ps. lxxii. 28.) Let us encourage ourselves by means of these considerations to do all for God, and to do it in the best possible manner. Let us pass in review all that we have to do to-day, and let us make a strong resolution to do it for God alone and with all possible perfection.

Resolutions and spiritual nosegay as above.

Tuesday after Septuagesima.

Summary of the Morrow's Meditation.

In order to conform ourselves to the Roman liturgy, which honors Jesus to-morrow in the Garden of Olives, we will meditate on this mystery, and we shall learn from it: 1st, to avoid sin; 2d, to sanctify the trials of life. We will then make the resolution: 1st, to recall Jesus Christ to ourselves in the Garden of Olives that we may be excited to compunction for our sins, whether in the evening at the examination of our conscience, or whilst preparing ourselves for confession; 2d, to foresee every morning the trials awaiting us during the day, in order to encourage ourselves to support them in a Christian manner and without complaint. Our spiritual nosegay shall be the words of Our Lord in the Garden of Olives: *"Father, if Thou wilt, remove this chalice from Me, but not My will but Thine be done"* (Luke xxii. 42).

Meditation for the Morning.

Let us transport ourselves in spirit to the Garden of Olives; let us there contemplate the Saviour of the world prostrate on the earth before the majesty of His Father, all bathed in the blood which flows out of His veins through sorrow for our sins and the love He bears us. Prostrate in spirit beside Him, as though to gather up the drops of this precious blood, let us adore Him in His anguish, and let us unite our sorrow with His, our tears with His blood (St. Bernard, *Serm.* iii. *in Nativ. Dom.* no. 4).

FIRST POINT.

Jesus in the Garden of Olives Teaches us to avoid Sin.

If we sin, it is because we do not fear sin enough : hence the facility with which we commit it ; it is because we do not detest it enough after having committed it : hence the defects in our contrition, hence the little fruit we derive from our confessions and perhaps the sacrilegious abuse of the sacrament of penance ; it is because we do not fight enough : hence our relapses. Now Jesus Christ, in the Garden of Olives, remedies these three great evils. 1st. He teaches us to fear sin. His holy soul saw sin, even venial sin, under quite a different aspect from what we do ; He understood what God is with His infinite love, and how dreadful it is to offend Him ; what hell is and how fearful it is to fall into it ; what heaven is and what unhappiness it is to lose it ; what a soul is and how much it is worth ; what grace is and how terrible it is to abuse it. And these great prospects, joined to the thought of all the sins of the world which He beheld before Him cast Him into mortal fear (Mark xiv. 33). Oh, if we were but to look in the same light at all the sins to which we attach so little importance: thoughts of self-love, ambitious desires and vanities, our calumnious words and ill-tempered ones, our culpable actions, our eternal search after comfort, our numerous omissions—how much more should we fear sin! 2d. The gravity of sin being thus understood, Jesus teaches us how to weep. His soul is sad even to death (Matt. xxvi. 38). He weeps with all His members, and sheds tears of blood (Luke xxii. 41). He sinks under the weight of grief, and it is necessary that an angel should come down from heaven to sustain Him; which recalls to us, at the same time, that our troubles are known to heaven, and that from heaven alone can come our true consolation (Ibid. 43). What a beautiful model of contrition! 3d. Jesus in the Garden of Olives teaches us how to fight against sin: He feels in Himself the greatest repugnance to the opprobrium, the torments, and the death which await Him; He struggles generously against this repugnance, and in spite of nature which murmurs, He rises and says, Arise, let us go where God calls us. Thus we must always serve God, spite of everything, do violence to our tastes and our aversions, and bend beneath the divine will, remembering that the revolts of nature against grace, and of the flesh against the spirit, can do us no harm as long as our will remains attached to God. Is it thus that we fear sin, that we detest it, that we struggle against it?

SECOND POINT.

Jesus in the Garden of Olives Teaches us to Sanctify the Trials of Life.

Jesus, in the midst of the severe trials which assailed Him, had recourse to three means: 1st. Prayer. He withdraws from His apostles and prays. Twice, through charity, He interrupts His prayer; twice He resumes it, and the more His anguish augments, the more He prays (Luke xxii. 43). In our troubles, we seek our consolation from creatures, we complain to them, and we derive nothing from it but more bitterness. Let us have recourse to God, and we shall be consoled (Ps. lxxvi. 3, 4). 2d. Jesus conforms His will to the will of God; He does not ask for His deliverance, but for the accomplishment of the adorable will (Matt. xxvi. 30). All sufferings, if they are not accompanied by this perfect abandonment to the divine good pleasure, not only lose their merit, but are converted into occasions of sin, by our impatience and our murmurings. Oh, why is it that we do not then say: "God is our Father; let Him afflict us, let Him strike us as much as He will, it can only be for our good." 3d. Jesus, in the midst of His trials, bears peacefully the indifference of His apostles. Twice He finds them asleep, although He had counselled them to watch and to pray; and without being angered, He only reproves them gently, excuses them, and exhorts them anew. What, you could not watch even one hour with Me? He said to them. "*Watch ye and pray, that ye enter not into temptation*" (Matt. xxvi. 41). Thus He teaches us never to make others suffer on account of our own troubles, not to complain of their indifference, and to excuse their wrongs towards us. Do we not often do the contrary?

Resolutions and spiritual nosegay as above.

Wednesday after Septuagesima.

Summary of the Morrow's Meditation.

We will resume to-morrow our meditations on the service of God, and we shall see: 1st, that all creatures invite us to serve God; 2d, that they offer us means for doing so. We will then make the resolution: 1st, not to please ourselves in creatures, but only in God, whom we must see in everything; 2d, that we must make use of all events here below as of so many steps to raise us to God by the adoration and love of His providence, of

His wisdom, of His patience, of His goodness. Our spiritual nosegay shall be this saying of the saints: "*How will that serve me for God, and for eternity?*"

Meditation for the Morning.

Let us adore the Creator making all the universe for man, and man for God. Let us bless Him for this admirable plan which binds together in one great whole all creation with the Creator, and let us repeat from the bottom of our hearts with a profound sentiment of gratitude and love: "All which exists exists only for me, and I exist only for God."

FIRST POINT.

All Creatures Invite us to Serve God.

The magnificence of paradise and the terrors of hell invite us to serve Him, the one through hope, and the other through fear. Mary, the angels, and the saints exhort us by their example. Heaven, which is above our heads, cries out to us by a thousand voices how little the earth is to him who looks on high. From the heavens which recount the glory of God I lower my eyes to the earth, and there all terrestrial things cry out equally to me after their manner: Do not stop at us; raise your hearts and your minds to Him who has created us for you. The house that you inhabit cries out to you: Love God who has given you the stones of which my walls are made. The furniture of which you make use says to you: Love God who has produced out of the earth the wood of which I am made. The clothing which covers you says: Love God who has furnished the materials from which I am woven. The meats on your table say to you: Love God who has created animals for you, vegetables and fruits which nourish you. The sun in shedding light upon you during the day, the moon and stars which direct you during the night, the fire in warming you, the water in refreshing you, the air in giving itself to be breathed by you, the flowers in rejoicing your eyes or in delighting you with their odors, cry out to you: "Lift up your hearts." The events of this lower world hold the same language; if they harmonize with your desires, they invite you to give thanks to God who has so ordered them. If they are contrary to your wishes, they invite you to profit by them, to grow in conformity with the will of God, in patience, in humility, in detachment, in holy desires for heaven, in prayer, that supreme consolation of afflicted souls, and thus to acquire a rich provision of merits by each act of patience, and another jewel for your crown by each victory over yourselves. Lastly, there is nothing, not even excepting sin, which ought not, after its manner, to raise our souls to God, by a humility full of confidence, by a fervent prayer like that of the publican, an energetic resolution to lead a better life in reparation for the past. Thus all turns to the good of those who love God, says St. Paul (Rom. viii. 23), even sin, adds St. Augustine, and we can and we ought also to add: even the sins of others, for they ought to be to us an opportunity for praising and imitating the patience of God, His goodness, His mercy, and to pray to Him for the conversion of poor sinners. Let us listen to this voice which issues from all parts of the creation to invite us to love and to serve God. The saints knew so well how to listen to it, and they profited by it in order to keep themselves always recollected in God and to encourage themselves to perfection.

SECOND POINT.

All Creatures Offer us the Means of Serving God.

Let us ask this secret of the saints. All creatures were to them as so many steps whereby they raised themselves to God, as so many mirrors in which were reflected to the eyes of faith the divine perfections, as so many focuses where their hearts were kindled with love for God by ever new flames; never stopping at created things, they passed from them to God as the first principle and the essential end of all that exists, and they raised themselves thereby every day from virtue to virtue. Let us imitate their example. When looking up to the skies let our hearts exclaim: Let us praise the Lord whose eternal mercy has created all these marvels for us (Ps. cxxxv. 1). When beholding the earth, its harvests, its meadows, its fruits, and its flowers, let us repeat the same cry of love: Praised be God, whose eternal mercy has made all these things for us. Witnesses of all the events which take place in this world, let us raise ourselves to the love of the Providence which directs all in a spirit full of wisdom and of goodness towards the elect. Even when beholding the sins committed on earth, let us raise ourselves to the love of that divine patience which bears so many outrages in silence. Happy he who thus makes use of everything for the purpose of elevating himself to God; but woe to him who, stopping at the creature, places therein his consolation and his happiness, who only looks at God as though only for a moment and as something which is an accessory thing in life. He misapprehends the destination of creatures; and, instead of raising himself by them to God, he makes of them instruments of sin and damnation which absorb all his senses, bind and corrupt his heart. Do I not give way to this kind of disorder?

Resolutions and spiritual nosegay as above.

Thursday after Septuagesima.

Summary of the Morrow's Meditation.

We will to-morrow consider in our meditation how we must serve God by the use of creatures, and we shall see how, in order to do so, we must use: 1st, things which are necessary to life; 2d, things which are not necessary. We will then make the resolution: 1st, to look upon all that happens to us and all that we have to do only as means for sanctifying ourselves; 2d, to desire nothing upon earth except the good pleasure of God, and in everything to prefer that which best leads us to our end, which is our salvation. Our spiritual nosegay shall be the same as that of yesterday: *"How will that serve me for God and for eternity?"*

Meditation for the Morning.

Let us adore God as the essential end of all things, as the adorable centre towards which all our thoughts and all our affections, all our projects and all our actions ought to converge. With this end in view let us render Him our homage.

FIRST POINT.
How to Use Things which are Necessary to Life.

Things which are necessary to life are sleep,

food, clothing, lodging, and the thousand little attentions which the body claims and the relaxations which our weak nature cannot do without. The worldly man places his happiness in low and miserable things. To sleep well, to eat well, to be well clothed and well lodged, to want for nothing, to amuse himself, and to do nothing—this, in his opinion, is supreme happiness, and he would willingly addict himself to it throughout eternity. But the true Christian thinks very differently. He feels himself humiliated at being the slave of so many necessities, condemned to act the corpse during the large portion of his existence which is given to sleep; to play the animal several times a day, in browsing like the beasts and assimilating his food like them; in hiding himself beneath clothing by a very legitimate shame of himself; to have for his lodgment and for comforts of the simplest kind so many needs which require the concurrence of the products of the earth and the fleece of animals and the arms of a thousand laborers; lastly, to do nothing during a notable part of his life, because otherwise his mind and his body would be overwhelmed with fatigue. If he satisfies these miserable necessities it is only in sighing and in observing the three rules of the saints: *take, thank, fear.* 1st. *Take*—take what is simply necessary, and nothing beyond that; take it, not to satisfy your body and to give it pleasure, but only from the desire to obey God, who so wills it. Take it in a spirit of humility and resignation, which submits to the necessity of its condition for the good pleasure of God, which is its only love. 2d. *Give thanks ;* in taking what is necessary thank God who gives it you in a better measure and under more advantageous conditions than to many others. If what is necessary is pleasant to your taste and to your senses, thank God who spoils you and treats you far better than you deserve; if you do not like it, still give thanks to God who gives you an opportunity of mortifying yourself and acquiring merits. 3d. *Fear*—be afraid of attaching your heart to the creature, be afraid of taking more than is necessary, and make God a generous sacrifice of what is superfluous. Be afraid lest the body should weigh down the soul; be afraid lest being too well satisfied it should revolt (Prov. xxix. 21). The observance of these holy rules will cause effeminacy, intemperance, the loss of time, pleasure taken solely because it is pleasure, to disappear; by means of them all ill-regulated habits will be restrained, all useless expenses will be retrenched, and all desires will be moderated. How many things there are which I can do without, we shall say to ourselves. But of them will be formed a rich foundation for charity, and the spirit of sacrifice, entering into the whole of our conduct, will raise the soul to holiness. Let us sigh, in the presence of God, that we have observed these holy rules so little, and let us propose henceforth to conform our conduct to them.

SECOND POINT.

How to Use Things which are not Necessary to Life.

By things not necessary is to be understood the more or less of amusements we might have for ourselves, the more or less of wealth which we might amass, the more or less of glory and reputation we might acquire, a certain kind of life, certain occupations, certain pastimes, such as games, visits, and other like things. The rule to be followed is to ask ourselves: What use can I make of that for God, for my eternity? If it be useful for these ends we must lovingly embrace it; if, on the contrary, it is hurtful, we must reject it with horror. If it be in itself neither hurtful nor advantageous, we must be indifferent to it, not desiring one thing more than another, health rather than sickness, riches rather than poverty, honor rather than contempt, a long life rather than a short one. The only legitimate preference in the use of created things is when they lead us more surely. Thus pious exercises well performed, profound devotion in prayer, habitual recollection, useful and moderate occupations, will bring me nearer to God; I will therefore apply myself to them. Sin, voluntary imperfections, dangerous occasions, too natural attachments, dissipation, an excessive desire to succeed, too great intercourse with the world, would turn me away from God; therefore I renounce them. If I meet in my path with indifferent things, I will make the sacrifice of them, and then they will become useful for my salvation. Alas! why have I not followed these rules? How much better should I have lived!

Resolutions and spiritual nosegay as above.

Friday after Septuagesima.

Summary of the Morrow's Meditation.

We will resume to-morrow, in three words, our meditations of this week; to be wholly God's is a duty, a glory, and a happiness. We will then make the resolution: 1st, to sacrifice to God the slightest attachments we are able to perceive in ourselves; 2d, often to repeat as an ejaculatory prayer: *All to God alone, all for God alone.* Our spiritual nosegay shall be the words of the Psalmist: "*What have I in heaven? and besides Thee what do I desire upon earth? For Thee my flesh and my heart hath fainted away*" (Ps. lxxii. 25, 26).

Friday after Septuagesima.

Meditation for the Morning.

Let us adore God as the supremely amiable Being, the only amiable One to whom it is just, honorable, and infinitely advantageous to attach ourselves. Yes, my God, to be Thine alone, Thine entirely, Thine always, is a duty, a glory, and a happiness.

FIRST POINT.

To be wholly God's is a Duty.

It is a duty of justice, because our whole being is from Him and of Him; it is a duty of gratitude, because we exist only by His benefits; it is a duty of conscience, since we cannot subtract anything from Him without our conscience telling us that we are doing wrong, and very wrong; it is a duty of honor and delicacy, since if God deigns, through compassion for our weakness, to leave many things in the rank of counsels, and not to press us in what He desires of us by making of it an express commandment, it ought to be an additional reason to make us generous in His service, and to do for Him as much as we can, and in the best possible manner; not that we ought to be troubled by our infidelities in what is not of precept, and allow ourselves to be thrown into scruples, but we ought to humble ourselves and be confounded in the presence of God, and repair past negligence by a greater fidelity.

SECOND POINT.

To be wholly God's is a Glory.

To give up a portion of our heart and of our time to serve creatures, the world, our passions, or bad inclinations, is a shame, an indignity, a degradation of the dignity of man and of the character of a Christian. True glory consists in raising all our intentions to God, without ever descending to the creature. Amongst men it is considered an honor to labor only for kings, and we ought to look upon it as an honor to labor only for God. We are too great to labor for the world; the world will pass, and we shall never pass away. We are the children of God; nay, we are even of the very same race as God (Acts xvii. 28), the friends, the confidants, the favorites of God; and being in so lofty a position, we ought not to lower ourselves to act for an end which is below God. Our vocation is to imitate God, to act in God, to live like God. Now, God proposes Himself alone as the end of all that He does. Our glory is to remain at this height, and not to descend to the little and low aims of the creature. What a shame it is for us to degrade ourselves when our destination is so sublime! Let us henceforth have more pride, and let us do nothing except for God.

THIRD POINT.

To be wholly God's is a Happiness.

If we have regard, in even the smallest degree, to any other end than God alone, we are unhappy; we fear, we desire, and a mere nothing which may be wanting to us poisons all the rest. Even when nothing is wanting, we feel that all is deception and vanity, trouble and bitterness, danger and a precipice. If, on the contrary, our hearts are wholly God's, we have peace, confidence, and happiness. We are well with God, because He becomes our friend as soon as ever we desire He should be, and a sure friend, who will not fail us except in proportion as we will to have it so. We are well with our neighbor, because the more we are God's, the more meek and humble, charitable, disinterested and equitable we are; that is to say, we become everything that gains the heart's esteem and affection. We are well with ourselves, because the heart which reposes fully in God is in its element; it finds therein life, happiness, an anticipated paradise. Let us reflect on how much pain we spare ourselves in being wholly God's, and how many subjects of affliction we create for ourselves by serving the creature.

Resolutions and spiritual nosegay as above.

Saturday after Septuagesima.

Summary of the Morrow's Meditation.

We will meditate to-morrow on the last words of the gospel of last Sunday: "*Many are called, but few are chosen;*" and we will consider: 1st, why there are so few of the elect; 2d, what we have to do in order to belong to this small number. We will then make the resolution: 1st, never to allow ourselves to be influenced by the example of the majority, but, on the contrary, to ask ourselves what the saints would have done in similar circumstances, what they would have said, what they would have thought, and thereby to rule our own conduct; 2d, to lay to heart the affair of our sanctification, and to pursue it with this sentiment in our heart: I will be a saint. We will retain as our spiritual nosegay the words of the gospel, "*Many are called, but few are chosen.*"

Meditation for the Morning.

Let us adore Our Saviour pronouncing those terrible words, "*Many are called, but few are chosen.*" Let us admire the sentiments of His heart at that thought—He who so loves men that He desires to save them all; and the prospect of so many souls who will abuse His mission, His passion, and His death, His sacraments, and all the means of salvation which He destines for them, all these things cut Him to the heart so as to make Him cry out in the Garden of Olives: "*My soul is sorrowful even unto death*" (Matt. xxvi. 38). Let us thank Him for the love He bears us, and let us promise to console Him by living the life of the elect.

FIRST POINT.

Why there are so Few of the Elect.

If there are so few of the elect it is not owing to God, it is owing to man; it is because: 1st, the majority do not think seriously of their salvation, and are determined not to think of it. To think of earthly things, well and good, it pleases them; but to think of what they will become when they leave this life, that is just what they cannot bear even to be made to think of. Similar to the laborers in our gospel, instead of working in the precious vineyard, the culture of which is confided to them, that is to say, their souls, they lose time in going from place to place; in talking about trifles and things that are taking place; they occupy themselves with nothing but earthly affairs, and they do not know how to raise their eyes to heaven. In order that they should be saved, it would be necessary that God should save them without their concurrence. Now St.

Augustine has said: God, who created us without our aid, will not save us without our concurrence. 2d. There are few of the elect, because many, even though they may think of it, dare not resolve for good and all to lead the life which saves. Cowardice stops them, holiness frightens them, the holiness which is so beautiful, which is the secret of happiness upon earth as it is in heaven. They limit themselves to saying: I would wish to be a saint, the elect of God, with the mental reservation that it will not cost me any sacrifice. They never say resolutely: It is decided, it is a fixed determination on my part. I will be a saint, and I shall be. With them it is but a feeble, cowardly will possessing no energy; one of those powerless and sterile desires of which hell is full, one of those half-wills of the idle who are killed by desires (Ps. xxi. 25), who will and will not, who say to themselves: *"There is a lion without, I shall be slain in the streets"* (Prov. xxii. 13); he will devour me. Now it is not thus that we save ourselves. In order to succeed, we must fix this project firmly in our head, take it to heart, pursue it diligently, saying to ourselves and often repeating it: I will be saved, I will, whatever it may cost me; I will, whatever may be said and whatever may be thought of me. Let us examine ourselves by these signs as to whether we are of the number of the elect.

SECOND POINT.

What we have to do in order to be of the Number of the Elect.

1st. We must avoid the wide road in which the majority walk, and follow the narrow way where but few are found. For since it is the majority who are lost, we cannot hope to save ourselves in living like the majority, but rather in living like the few; that is to say, in not allowing ourselves to be carried away by the habits and customs of the world, and never losing sight of the fact that even amongst Catholics there are but few Christians, whether in the town or in the country. 2d. We must always have present to our mind the signs by which the wide road is distinguished from the narrow, in order not to confound the one with the other. In practice, the wide road is recognized by this sign, that men will not put themselves out of their way, but will live at ease and without constraint; consequently they think that it is sufficient to avoid gross vices and not to do harm to any one. They do not in the least aspire to be saints, but leave others to do so; it is enough to live as do the common herd. They do the least which is possible for their salvation, choosing in religion just what pleases them and leaving the remainder on one side. They certainly propose to live better later on, but the moment never arrives. The narrow way, on the contrary, is recognized by these signs, that men fight therein against their inclinations and their passions, above all against their besetting sin ; that they perform their duty, cost what it may ; that they renounce themselves ; that they mortify themselves ; that they bear their cross ; that they watch over their hearts and over their senses. That seems hard, but the practice is full of sweetness. 3d. We must labor for our salvation with courage and confidence. Why should I not do what so many others have done before me? The man who has a resolute will can do everything with the help of grace, which is never refused to him who asks for it. The soldier to fulfil his duty, the merchant to make his fortune, the laborer to gain his livelihood, impose many more cares and sacrifices upon themselves than are demanded by religion. That the man can be saved who wills to be, is an article of faith. Let us examine ourselves upon these principles as to the way in which we walk. Are we not content to live like the great majority, to follow our own ease without constraint, to take in religion just what pleases us, to leave the rest on one side ? Do we aspire to imitate the saints and the little number of the elect?

Resolutions and spiritual nosegay as above.

Sexagesima Sunday.

The Gospel according to St. Luke, viii. 4–15.

"And when a very great multitude was gathered together and hastened out of the cities unto Him, He spoke by a similitude. The sower went out to sow his seed. And as he sowed some fell by the wayside, and it was trodden down, and the fowls of the air devoured it. And other some fell upon a rock; and as soon as it was sprung up, it withered away, because it had no moisture. And other some fell among thorns, and the thorns growing up with it choked it. And other some fell upon good ground; and, being sprung up, yielded fruit a hundred fold. Saying these things, He cried out: He that hath ears to hear, let him hear. And His disciples asked Him what this parable might be, to whom He said: To you it is given to know the mystery of the kingdom of God; but to the rest in parables, that seeing they may not see, and hearing may not understand. Now the parable is this: The seed is the word of God. And they by the wayside are they that hear, then the devil cometh and taketh the word out of their heart, lest, believing, they should be saved. Now they upon the rock are they who, when they hear, receive the word with joy: and these have no roots; for they believe for a while, and in time of temptation they fall away. And that which fell among thorns are they who have heard, and going their way are choked with the cares and riches and pleasures of this life, and yield no fruit. But that on the good ground are they who in a good and very good heart, hearing the word, keep it, and bring forth fruit in patience."

Summary of the Morrow's Meditation.

We will meditate during the whole of the week on which we are about to enter upon the gospel of to-morrow, which treats of the word of God; and in our next meditation we will consider: 1st, the excellence of this divine word; 2d, the ways in which God points it out to us. We will then make the resolution: 1st, to receive with great respect and lively gratitude the word of God, in what manner soever it may reach us, whether by public instructions, or by good books, or by good thoughts; 2d, after having listened to the divine word, to preserve it as a treasure in the bottom of our hearts and to conform our conduct to it. Our spiritual nosegay shall be the passage in the gospel which presents the Blessed Virgin to us as a model in this as well as everything else: "*Mary kept all these words, pondering them in her heart*" (Luke ii. 19).

Meditation for the Morning.

Let us adore Jesus Christ teaching us, by the parable of the seed, that which is most useful for us to know with regard to the word of God. Let us bless Him for this condescension, which, in order to bring within reach of the most humble minds the most sublime truths, lowers the elevation of its thoughts down to the most simple and most common comparisons.

FIRST POINT.

Excellence of the Word of God.

St. Ambrose, after having quoted the passage of the psalm, "*Thy Word is exceedingly refined*" (Ps. cxviii. 140), adds this beautiful commentary: Fire purifies by separating gold from rust; it lights, it heats. In the same way, the word of God purifies souls, enlightens the intelligence, warms the heart. 1st. It purifies. It renders the proud humble, the vain modest, the unchaste pure, the miser generous. How many sinners owe their conversion to it! How many lukewarm owe to it a better life! (Ps. cxviii. 8.) 2d. It enlightens. On one side it reveals to the soul the falsity of earthly pleasures, the nothingness of riches, the illusion of glory, and it rectifies the false judgments of our blind passions and of our corrupted senses; on the other side, it makes the pure light of faith shine before our eyes like the pillar in the desert—it guides our steps in the path of life (Ps. cxviii. 105); and a few pages of the catechism teach more to man of those things which it is most important for him to know than all the books of human wisdom (Ps. cxviii. 99).

3d. It inflames. It kindles the fire of life in souls which are dead in sin, and makes charity burn where once passion burnt. Carried by an Augustine into England, by a Boniface into Germany, by a Xavier into India, by a St. Dominic, a Vincent Ferrer, a Thomas of Villanova, a Borromeo, a St. Francis de Sales, into divers parts of the earth; by zealous pastors into thousands of Catholic parishes—everywhere it kindles the sacred fire in the heart. And here what reproaches have I not to address to myself? Through my fault, the holy word has not purified me; it has not freed my virtue from all alloy; the rust of a thousand little passions is still eating into my soul. What attachments soil me! What ill-regulated affections share my heart! Through my fault, the holy word has not enlightened me. Blinded by a life of habit and of routine, entirely human and natural, I cannot derive my judgment and my manner of looking at all things from my faith. Lastly, through my fault, it has not inflamed me; I am lukewarm, if not cold, in the **service** of God.

SECOND POINT.

The different Ways in which God speaks to us.

God, in His infinite goodness, has multiplied the channels through which to make His word reach our hearts: 1st, by oral instruction, whether given in Christian pulpits, or at the holy tribunal, or in the administration of the sacraments, or in the counsels which His providence gives us by divers organs. How great goodness is there in this conduct of God towards us, and how much more He favors us than He does so many millions of men who are spread over the globe! He speaks to us, 2d, by the holy books and all the pious works which we can read. This kind of reading has converted thousands of sinners, and every day it feeds and perfects piety in souls. He speaks to us, 3d, by good thoughts, pious movements, salutary remorse, warnings, and the light His grace sheds on us; sometimes in prayer, at Holy Communion, in visits to the Blessed Sacrament; sometimes at even the most unexpected moments. Happy the souls who are sufficiently recollected to listen to this voice, and generous enough to obey it. He speaks to us, 4th, by the good examples He places before our eyes. Each good example is a sermon which teaches us, here, meekness, patience, devotion; there, reverence in church, assiduity in assisting at its offices, frequentation of the sacraments. What fruit do we derive from so many means of salvation?

Resolutions and spiritual nosegay as above.

Monday after Sexagesima.

Summary of the Morrow's Meditation.

We will meditate to-morrow upon three obstacles which prevent the word of God from producing fruit in the soul. Our Lord has pointed them out to us in the three kinds of ground where the seed falls. The first obstacle is worldliness, figured by the beaten road, open to all passers-by; the second is cowardice, figured by the inert, stony ground, which does not allow vegetation to take root; the third is attachments, figured by the stones which cover the earth. We will then make the resolution: 1st, throughout the day to preserve a spirit of more than ordinary recollection, in order to profit by the good sentiments which the Spirit of God may suggest to us; 2d, not to refuse any sacrifice to grace. Our spiritual nosegay shall be the words of the Apostle: "*The earth that drinketh in the rain which cometh often upon it, and bringeth forth herbs meet for them by whom it is tilled, receiveth blessing from God; but that which bringeth forth thorns and briars is reprobate, and very near unto a curse*" (Heb. vi. 7, 8).

Meditation for the Morning.

Let us render to Jesus Christ our homage of adoration, of praise, and of thanksgiving for His goodness in pointing out the obstacles calculated to render His divine word sterile in us. Oh, how precious is this lesson! May we understand it well, and profit abundantly by it.

FIRST POINT.

The First Obstacle to the Word of God is Worldliness.

The soul which is worldly is indeed the beaten road, open to all passers-by, where all the world comes and goes, passes and repasses, treading under foot the divine seed which is afterwards devoured by the birds of heaven (Luke viii. 5). Traversed in every direction by a thousand vain and useless thoughts, full of the world and its affairs, but empty of the interior spirit of recollection and union with God, the poor soul is always occupied with what is passing around it, and hardly ever occupied with itself. The past, the present, the future, all absorb it; and in this deplorable state there is no means of preventing the divine seed from being on the one side trodden under foot by all the vain thoughts which ceaselessly pass and repass, and on the other from being carried away by the birds of heaven, that is to say, the vain imaginations which also traverse

Obstacles to the Word of God.

the atmosphere. Such a soul may still make laudable resolutions to pray, to read spiritual books, and that will be the word of God ready to take root; but it will not watch over itself; ideas from without will arrive and will cast themselves upon the seed; it will be recollected no longer, worldliness will soon have ruined everything. Is not this our history?

SECOND POINT.

The Second Obstacle to the Word of God is Cowardice.

Another portion of the seed, said Jesus Christ, falls upon stony ground, germinates at first without difficulty, but meeting with stones, dries up and dies (Luke viii. 6). By this is meant, said the Saviour, those who receive the divine word without repugnance and even with joy, who love to hear about God and religion, to read pious books, but who on being exposed to the trial of a sacrifice or of a difficulty lose courage and withdraw (Ibid. 13). Cowardice in the service of God, this then is the stone which is at the bottom of the heart, which dries up the divine grain, and prevents it from growing. As long as there is no sacrifice to make, all is well, the grain germinates and springs up in good sentiments and holy affections; but as soon as a difficulty to be surmounted presents itself, a temptation to be vanquished, a sacrifice to be made, there is a stop page. The stone is there: it is cowardice; the seed cannot penetrate it, it dries up and dies. The soul desires to love God, but on condition that it costs nothing; it desires indeed to be saved, but without doing any violence to itself. It admires the saints, but without imitating them; it has not the courage to do so; the stone is there: it is cowardice. It reads indeed in the gospel that it must renounce itself and take up the cross; but these words hardly touch its surface, and it does neither less nor more, because the stone is below, which hinders them from sinking into it: it is cowardice. Oh, who will remove this stone (Mark xvi. 3) that the new man may come forth and that the word may fructify.

THIRD POINT.

The Third Obstacle to the Word of God is Attachments.

This is what Jesus Christ points out to us by the thorns and briars which stifle the seed (Luke viii. 7). The earth is good in itself, for the abundance of briars proves the fertility of the soil; there is a certain amount of courage and energy in the soul; it welcomes the virtues recommended by the divine word and decides to practise them; but in the midst of these good resolutions it allows to grow and be rooted in it certain attachments which it will not break off; attachments to a comfortable and sensual life, to pleasure, to money, to glory, to reputation, to its own will, to its character, to its own opinion and its own views. All these attachments grow and are developed, they cover the good resolutions which have been made, stifle them in the bud, and thus render the divine word sterile at the very moment when it was about to form itself into ears of corn. Is not this our history?

Resolutions and spiritual nosegay as above.

Tuesday after Sexagesima.

Summary of the Morrow's Meditation.

In order to conform ourselves to the Roman liturgy, which to-morrow sets before the faithful the Passion of the Saviour, we will consider: 1st, that devotion to the Passion is a duty of the heart; 2d, that everything in religion prescribes this duty to us. We will then make the resolution: 1st, to have a crucifix always on the wall of our chamber, another upon our writing-table, and the third on our breast; 2d, to salute the crucifix each time that we enter into our chamber, or wherever we perceive it. We will retain as our spiritual nosegay the words of the Apostle: *"Think diligently upon Him that endured such opposition from sinners against Himself"* (Heb. xii. 3).

Tuesday after Sexagesima.

Meditation for the Morning.

Let us adore Jesus crucified, the great object of the devotion of all Christians. Let us unite ourselves with those who place all their happiness in meditating upon the crucifix and by means of this meditation gain the most sublime virtues.

FIRST POINT.

Devotion to the Passion is a Duty of the Heart.

It would be to have no heart if we were to forget so great a benefit, to look with indifference at the crucifix, that most splendid trophy of the charity of a God, that admirable invention of the entrails of His mercy (Luke i. 78), to which we owe everything, the adoption of children of God, grace during life, and glory in eternity. If a friend had given his life for us, and had died in our stead in ignominy and torments, we should remember it until our last sigh, we should recall with emotion all the circumstances of his agony, and we should kiss with tears of affection the picture representing him suffering and dying for us. How much more should the love of Jesus crucified be impressed on our hearts and make us live and breathe only for Him! (II. Cor. v. 14, 15.) For upon the cross it is not for His friends that Jesus dies, but for those who had even become His enemies (Rom. v. 8, 9). It is written: "*Forget not the kindness of thy surety: for he hath given his life for thee*" (Ecclus. xxix. 20). And this generous friend who has answered and paid for us, all unworthy as we were of such love, who is he, if it be not the Divine Crucified Jesus? Therefore, says St. Augustine, if he who forgets the blessing of creation merits hell, he merits a thousand hells who forgets the blessing of redemption. Nevertheless, how many are there who scarcely ever think of it! Just because they have the crucifix constantly before their eyes, they become insensible to it, and just because they have the spectacle of love always in view, they become ungrateful; that was the great sorrow of St. Francis of Assisi, that illustrious lover of the crucified Jesus. Day and night he shed tears over the ingratitude of men with regard to the cross of the Saviour, and when attempts were made to console him, "No," he replied, "all my life shall be inconsolable that my Saviour having so loved men, men should, nevertheless, love Him so little." Are we not of the number of those for whom the holy patriarch wept? What is our love for the crucifix? Do we ourselves wear it? Do we kiss it often? Do we look at it lovingly?

SECOND POINT.

Everything in Religion Preaches Devotion to the Passion of the Saviour.

The holy Mass, which is the principal act of religion, is nothing more than the reproduction of the sacrifice of Calvary. Consecrate and eat the Eucharist, said Jesus Christ to His apostles, in remembrance of Me, that is to say, according to the commentary of St. Paul, in memory of My death, in honor of My cross (I. Cor. xi. 26). How admirable! Jesus Christ, wishing to inspire us all with a constant devotion for His cross, institutes, to be a memorial of it, not a temporary sacrament like the other sacraments, but a sacrament which alone has the privilege of being permanent, a sacrament which we possess day and night in the holy tabernacle, where this adorable Saviour lives continually in the state of a victim, preserving all the wounds upon His body, and showing them ceaselessly to us, in order that we may never lose the memory of them. Oh, who is there that would not respond to the desire of a God who conjures us not to forget Him, and who conjures us by a sacrament of so great price, by His last words, which the whole world looks upon as sacred : "*Do this for a commemoration of Me*" (Luke xxii. 19). All that we see in church equally preaches to us devotion to the Passion : the cross is above the tabernacle, as being the place where it is most apparent, and where it first strikes the eye; it is carried in processions; it surmounts the summit of churches; it is represented upon the sacred vestments; the august sign of it is made in every ceremony; one day in each week, Friday, and feasts at different epochs are consecrated to it; a particular season of the year, the fortnight before Easter, is wholly given up to it; the way of the cross attracts the devotion of the faithful everywhere and at all seasons: so greatly does the worship of Jesus crucified enter into the essence of Christianity. And therefore it is that the saints, in whom is found the plenitude of the Christian spirit, have made the cross the most habitual object of their piety. St. Paul glorified himself in the cross alone; he would know nothing save the cross; he lived always attached to the cross (Gal. ii. 19). St. Augustine tells us that he fed his soul with meditation on the cross. St. Francis of Assisi would not allow his followers to have any other object of meditation than the cross which he had placed in the place of reunion of his confraternity. St. Bonaventure lived only in the wounds of the Saviour. "It is there," he said, "where I watch, where I take my repose, where I read, where I converse, where I will always be." It seems, remarks St. Francis de Sales, as if when this great doctor wrote down the heavenly effusions of his soul, he had no other paper than the cross, no other pen than the lance which had pierced his Master's side, no other ink than His precious blood. Oh, how far are we from possessing these sentiments of the saints with regard to the Passion of the Saviour! Let us reanimate our faith; let us rekindle our love for Jesus crucified!

Resolutions and spiritual nosegay as above.

Wednesday after Sexagesima.

Summary of the Morrow's Meditation.

We shall resume to-morrow our meditations upon the word of God, and we shall consider: 1st, the respect, 2d, the attention we owe to the divine word. We will then make the resolution: 1st, to listen to and to read the word of God with the same respect as though God Himself were speaking to us, and not to criticise sermons; 2d, to seek in instructions not what amuses the mind, but what changes the heart, and also to make practical resolutions derived from it. Our spiritual nosegay shall be the words of St. Augustine: *"To listen with a careless ear to the word of God is the same crime as to allow the sacred Host to fall to the ground through carelessness."*

Meditation for the Morning.

Let us render our accustomed homage to Jesus Christ, and let us listen to Him declaring that the good ground in which the grain brings forth a hundred fold is the good heart which respects His word—the very good heart which listens to it attentively, that it may put it in practice (Luke viii. 15). Let us thank Him for this lesson, and let us beg Him to make it enter deeply into our hearts.

FIRST POINT.

The Respect due to the Word of God.

As soon as God, that infinitely great and elevated Being, deigns to abase Himself so low as to speak to man, a creature so base and so miserable, is it not evident that no respect is great enough for a word which descends from on high, no veneration profound enough, and that every word which emanates from so august a source ought to be received with the whole submission of the mind and all the obedience of the will? If we had heard upon Sinai God speaking to the children of Israel in the midst of thunder and lightning; or if living at the time of Jesus Christ, we had assisted at one of His discourses, we should have looked upon it as a great crime to have lent to the divine word nothing but an indifferent ear. But is, then, this word less worthy of respect if it be contained in the sacred pages of our divine books, or in the pulpit, where our ears listen to it? Man, who comments on it, may mingle with it his weakness and his ignorance, but it does not the less remain the word of God; and as the Word Incarnate was not less adorable in the poor swaddling-clothes of His infancy than in the splendor of the saints, so the word of God is not less venerable beneath the rags with which the ignorance of men envelop it than beneath the magnificent garments with which genius can clothe it. When an ambassador speaks we pay less attention to the more or less elegant style of his discourse than to the majesty of the prince in whose name he speaks. In the same way, in the minister of the divine word we ought to see only the ambassador of God, the lieutenant of Jesus Christ, who speaks to us by his mouth (II. Cor. v. 20; ii. 17). And looked at in this manner, the word of God has no less a claim upon our respect than the very body of Jesus Christ, as St. Augustine says; we ought to gather up all the particles of it as religiously as the priest gathers up the particles of the holy Host upon the sacred paten; and the negligence which allows them to be lost is not less culpable than that which would allow the body of the Saviour to fall to the ground. The reason is that Christ does not love truth less than He loves His own body; He seems to love it even more, since for it He sacrificed His body. He willed that it should be immortal upon earth, and His body He delivered to death. Is it thus that we respect the divine word? What reproaches have we not to address to ourselves on this subject! Let us humble and correct ourselves.

SECOND POINT.

The Attention with which we ought to Listen to the Word of God.

We listen to worldly matters, to frivolous stories, with a vivacity of attention which does not lose the smallest portion of them; we read letters from our relatives and friends with an interest which impresses them on our memory. Why, then, when the divine word gives us tidings of heaven, our country, and instructions on the manner of arriving there; why, when we have in hand the sacred books which are as so many letters which God sends us, do we become careless and inattentive? Wherefore have not these things the same attraction for us? Jesus says, Listen to My word in the depths of your hearts (Luke viii. 15). What does that mean? It means not only with the understanding, where the eyes see only the outward appearances, where the ears hear only the sound, the memory preserves nothing more than envelopes; but listen in that secret part of the heart which adores truth, appreciates and preserves it; listen, not in that portion of yourselves where periods are measured, but where morals are regulated; not in the place where beautiful thoughts are enjoyed, but in the place where good desires are produced; not in the place where opinions are formed, but in that where resolutions are taken; and if there be some place still more profound and more retired, where the council of the heart is held, where all its designs are decided, where the impetus is given to all its movements, it is there that we must withdraw and render ourselves attentive to the word of Jesus Christ (Luke xxi. 14). It was thus that the Blessed Virgin listened (Luke ii. 19); it was thus that St. Mary Magdalene listened at the feet of Jesus (Luke x. 39); whilst we, we are far from observing the same religious attention to the divine word, whether when our eyes read it or our ears listen to it. Let us ask pardon for the past, and resolve to do better for the time to come.

Resolutions and spiritual nosegay as above.

Thursday after Sexagesima.

Summary of the Morrow's Meditation.

We will meditate to-morrow on the means for profiting by the word of God, and we shall see:

1st, that we must listen to it with faith ; 2d, we must make the application of it to ourselves ; 3d, we must deduce practical resolutions from it. We will then make the resolution : 1st, at each instruction or at each reading to represent to ourselves God Himself who instructs us and to apply what He says to ourselves ; 2d, to draw from each instruction or reading practical resolutions for the reformation of our life. Our spiritual nosegay shall be the advice of St. James : *" Be ye doers of the word, and not hearers only : deceiving your own selves "* (James i. 22).

Meditation for the Morning.

Let us adore Our Saviour Jesus Christ teaching His apostles the science of salvation ; let us admire the faith and silence with which they listened to their adorable Master, and how they applied to themselves the words He said and put them in practice. Let us ask for a share in their dispositions and their graces, in order that we may rightly profit by the word of God.

FIRST POINT.
We must Listen with Faith to the Word of God.

Too often we listen to the divine word as though it were the word of man, as though it were a secular discourse, through curiosity, to appreciate the merit of it, or with carelessness as though it were something indifferent. This is an error which results in fatal consequences. At the sound of this holy word we must say to ourselves: It is not a man, it is God who speaks to me, the same God who will one day be my Judge. The day will come when He will ask an account from me of all that I hear. His word never returns to Him empty; it either bears fruits of benediction, if we profit by it, or fruits of condemnation, if we leave it sterile. It is God who speaks to me, God with His sovereign authority. I ought then to listen to Him religiously, with perfect docility of mind and of heart, without permitting myself to find anything in it to censure, without prejudice, or rather sacrificing to it all my prejudices, if anything of the kind should present itself to my mind. It is God who speaks to me, and who speaks to me for my good, to teach me the way to heaven and urge me to walk in it (Luke i. 77). I ought, therefore, to listen with this design in view: to seek nothing in the divine word but the means of becoming better, and to pray to God to enlighten me, to touch me, to make me put in practice His holy counsels (I. Kings iii. 9). Is it thus that I listen to the word of God? Do I see in him who announces it to me the God whom he represents, without paying attention to what is human in him, his style, his gestures, his voice, the whole of his exterior? Do I listen to him as though God Himself were there and had come down from heaven to instruct me?

SECOND POINT.
We must Apply the Word of God to ourselves.

The word which we do not apply to ourselves is like the arrow which flies above the head of the enemy without touching him; it is the seed carried away by the wind, which, not sinking into the earth, cannot germinate there, or produce fruit. This is why so many sermons and readings have been useless to me. I have said to myself: "This applies excellently well to such and such a person," and scarcely ever have I said to myself: "This applies exactly to me. This is indeed the faithful portrait of my conscience, of my character, of the state of my soul." If, instead of reasoning thus, I had, seriously reflecting upon myself, opened the door of my heart to the divine word, it would have revealed to me, through the many foldings of my soul, hidden passions, secret attachments, and the voluntary imperfections which exist in me. For the word of God, says St. Paul, is living and efficacious; more penetrating than a two-edged sword, it pierces the very marrow of the heart, dividing the soul and spirit, that it may discern its hidden miseries (Heb. iv. 12). Let us examine whether we are faithful in applying to ourselves the lessons that we hear and our readings of the holy word.

THIRD POINT.

From all the Instructions we Receive we must draw Practical Lessons tending to the Reformation of our Life.

"*Be ye doers of the word,*" says St. James, "*and not hearers only : deceiving your own selves*" (James i. 22); it would be to imitate the man who looks for a moment at his face in a faithful mirror, then goes away and forgets it (Ibid. 24). Of what use is it to see our miseries in the faithful mirror of the divine word, if, forgetting what we have seen, we do nothing to correct ourselves and take no resolution tended to render us better? We profit by the divine word only so far as we have patience to reform ourselves and to conquer ourselves, like those of whom it is said that they bring forth fruit by dint of patience (Luke viii. 15). Can that be said of us? Let us examine our conscience and correct ourselves.

Resolutions and spiritual nosegay as above.

Friday after Sexagesima.

Summary of the Morrow's Meditation.

We will meditate to-morrow : 1st, on the reasons which ought to induce us to read and meditate upon the Holy Scriptures; 2d, on the manner of doing this well. We will then make the resolution: 1st, not to allow a single day to pass without reading at least one chapter of the Holy Scriptures; 2d, from time to time to read at least an abridgment of the edifying history contained in the pages of the Old Testament. Our spiritual nosegay shall be the verse of the Psalm: "*How sweet are Thy words to my palate, more than honey to my mouth*" (Ps. cxviii. 103).

Meditation for the Morning.

Let us adore the Holy Spirit inspiring the sacred writers, leading their hand, and by their pen leaving to all ages the storehouse of holy truths which are the riches, the treasure, and the consolation of Christians. Let us thank God for so precious a grace.

FIRST POINT.

The Reasons which ought to induce us to Read and Meditate upon the Holy Scriptures.

What are the Holy Scriptures? They are a letter which God sends us, after having dictated it for us to the sacred writers. Now, if a monarch wrote a letter to one of his subjects, and if, after having received it, he did not care to open and to read it, would he not be guilty of a great want of respect? With still greater reason we should be inexcusable if we neglected the reading of this divine book, in which God, in order to lower Himself to our level, hiding Himself beneath the surface of the letter, speaks to us as really as He speaks to the angels and saints in heaven. If an interview with princes and kings is esteemed to be a very precious privilege what ought not to be the reading of a book in which God speaks to us? Is His word less venerable when the eye reads it than when the ear listens to it? In the one case as in the other, is it not always His word? When we pray, says St. Ambrose, we speak to God, and when we read the Holy Scriptures it is God who speaks to us. Whence this great doctor inquires: Wherefore do you not employ every moment in reading the Holy Scriptures, that is to say, in conversing with Jesus Christ? (St. Ambrose, *de Officiis*, lib. iii., c. xx.) We have all kinds of interests to consult in this employment, for of all books it is the most useful. Holy Writ is a treasure above all treasures (Ps. cxviii. 72, 162). It is there that the infidel is converted (Ps. xviii. 8); the reading of the gospel has converted thousands of men, and a few lines of the apostle St. Paul sufficed to fix the irresolution of St. Augustine. It is there that the afflicted soul finds its consolation. The Machabees in the midst of persecution consoled themselves by the reading of the Holy Scriptures (I. Mach. xii. 9), and St. Paul invited the Romans to seek for consolation in the Scriptures (Rom. xv. 4). It is there that the tempted soul finds a sure weapon against sin (Ps. cxviii. 11, 92), the soul that is a prey to weariness and disgust a delicious suavity (Ps. cxviii. 111), the soul in the midst of darkness, a light to direct it (Ps. cxviii. 105), the cold or tepid soul, a fire which warms it (Ps. cxviii. 140), the soul disgusted with the world, a sweet repose (Ps. cxviii. 85). Therefore the Psalmist lovingly meditated every day upon the Holy Scriptures (Ps. cxviii. 97). Do we act in like manner? Let us condemn ourselves and be converted.

SECOND POINT.
The Manner of Reading the Holy Scriptures.

Although all the books of the Bible are useful to read, there are some amongst them the reading of which is more profitable and ought, consequently, to be more familiar to us; such are, in the New Testament, the gospels, and in the gospels, the Sermon on the Mount, and the discourse after the last supper; the Acts of the Apostles, the Epistles of St. Paul to the faithful of Corinth and of Ephesus, to the Philippians, the Colossians and the Hebrews; the Epistles of St. James and St. Peter; in the Old Testament the book of Wisdom, of Ecclesiasticus, of Ruth, of Tobias, of Esther, of Judith, and of the Machabees, certain portions of the Pentateuch, of the Judges and of Kings. In order that this reading should be profitable, we must read, not from curiosity or the desire to learn something new, but, 1st, with a very pure intention to derive from it lessons and examples of virtue which may make us better; 2d, with a spirit of obedience to the Church, which alone is the infallible interpreter of Holy Writ, and in order to do so it is essential for us to use a correct translation, accompanied with short commentaries; 3d, in the presence of God, as though God Himself were there, instructing us, says St. Basil; and therefore we must ask of Him by frequent aspirations to give us understanding, and to make us feel what we read (Ps. cxviii. 18); 4th, in pausing at any passage which may touch us as long as we are touched, in order to enjoy the things of God, and to allow the Holy Spirit leisure to act within us. Having thus performed our reading, we deduct from it practical resolutions tending to render us better. Have we followed these rules? If we have not followed them up till now, let us make a resolution to do so in future.

Resolutions and spiritual nosegay as above.

Saturday after Sexagesima.

Summary of the Morrow's Meditation.

We will meditate to-morrow upon spiritual reading, and we shall see: 1st, what is the excellence of it; 2d, how we must perform it. We will then make the resolution: 1st, to be exact in reading every day some spiritual book, and in order that we may do so, to fix the precise time in our rule of life; 2d, to derive from this reading, as we do in meditation, practical resolutions suited to render us better. Our spiritual nosegay shall be the words which history recounts of St. Ephrem: *"He reproduced in his actions the pages he had read"* (Ennodius, *in Vita S. Ephrem*).

Meditation for the Morning.

Let us adore the providence of God, who by means of spiritual books makes us enjoy the conversations of the saints, their counsels, and their experience, and thus gives us a share in the lights and the good sentiments with which the Holy Spirit favored them during their lives. Let us thank His goodness for the great advantages which this kind of reading offers us.

FIRST POINT.

The Excellence of Spiritual Reading.

All the saints have esteemed this practice as one of the most important in regard to the spiritual life. St. Paul exhorts Timothy to it. "*Till I come, attend unto reading, to exhortation, and to doctrine*" (I. Tim. iv. 13), he writes to him. St. Jerome recommends the same practice to Nepotian. "*Every day,*" he says, "*be faithful to the reading of some good book.*" In order to understand the importance of this advice, let us represent to ourselves a great monarch, who, wishing to introduce to his court one of his subjects who is a stranger to the customs of this new abode, deputes to instruct him some of his principal officers. With what zeal ought not the favored subject to listen to the advice of the envoys of the prince! Now, spiritual books are precisely these envoys whom God sends us to correct in us what is incompatible with the life of heaven, and to render us worthy to take our place among the angels and saints. The reading of their writings makes these men of God, so worthy of all our respect, live again, in order to instruct us. We need not envy the happiness of their contemporaries; they speak to us by their writings as though they lived with us, and by a happy charm, by a divine enchantment, we enjoy them and their precious intercourse, with this difference: that in some ways we gain more by reading their writings than by listening to their discourses. Because, 1st, sermons are forgotten and cannot be recalled to memory, whilst, on the other hand, we can always have recourse to the books which contain what we have read; 2d, preaching passes away like a flash of lightning, and it is difficult to meditate upon it; but books remain before our eyes as long as we like, and we have leisure to meditate upon them in all their parts, to incorporate them, and to change them into our own proper substance; 3d, in listening to preaching, we only pass quickly before the sacred fire, and we have hardly time to warm ourselves at it; when reading, we remain as long as we like before the divine fire, we can be penetrated and heated by it as long as we like; 4th, when listening to sermons we often apply to others what we hear, without thinking of applying it to ourselves; when we are reading, on the contrary, alone with our book, we apply holy truths much better to ourselves; 5th, a book descends into practical details which are not always in keeping with the more elevated style of a sermon. Hence it is that spiritual reading has changed so many sinners into saints. Witness the two courtiers of the Emperor Theodosius, who were converted by reading the life of St. Antony; and St. Ignatius, who was converted by reading the Lives of the Saints. Witness, lastly, the experience of every day: a good reading well performed raises up the discouraged soul, consoles the desolate soul, encourages the languishing soul, fortifies the weak, makes the frivolous recollected, heats the cold and tepid, perfects the just, to such a degree that whoever is faithful to his daily reading sustains himself and advances in piety, and whoever is negligent in it goes back. Is it thus that we esteem spiritual reading? Are we faithful to it every day? Have we a time set apart for this exercise?

SECOND POINT.
The Manner of Reading Properly Spiritual Books.

In order that this kind of reading should produce all its fruit in the soul, the book must, 1st, be well chosen. It ought not to be a scientific book, nor a book difficult to understand, nor an amusing and curious book, because that would distract the mind, and would dry up the heart. It ought to be a pious book, exact and solid in doctrine, suited to show us, as in a mirror, our duties and our failings. Such are: Rodriguez' *Christian Perfection*, *The Imitation of Jesus Christ*, *The Spiritual Combat;* the various works of Father de Grenade; St. Francis de Sales' *Introduction to a Devout Life*, his *Spiritual Colloquies*, his *True and Solid Piety;* Lombez' *Interior Peace*. Such are also the Lives of the Saints, of St. Vincent de Paul, of St. Francis de Sales, of St. Ignatius, of St. Francis Xavier, of St. Aloysius. Are these our books for spiritual reading? 2d. The book being thus chosen, we must not read it either from curiosity, for that would be to fail in the object we have in view in reading it, and would close our heart to the operations of grace; nor for the sake of the beauties of its style, for that would be to imitate the fool who eats the leaves of the tree and leaves its fruit. We must read with the object and with the great desire to become better, to love and serve God better, and better to fulfil all our duties. Is this the object of our reading? 3d. The intention being thus laid down, we must, before beginning to read, recollect ourselves before God in order to dispose ourselves to listen to His voice, and we must pray to Him to speak to our heart (I. Kings i. 9, 10). Then we read quietly, appreciating and weighing what we read; pausing at passages where we are touched and as long as we are touched; applying what we read to ourselves, and deducting from it resolutions to reform such and such a particular defect in our life, according to what we read. By reading thus, we read little, but we read well, because we reflect much. Is it thus that we perform our spiritual readings?

Resolutions and spiritual nosegay as above.

Quinquagesima Sunday.

The Gospel according to St. Luke, xviii. 31-43.

"Then Jesus took unto Him the twelve and said to them : Behold, we go up to Jerusalem, and all things shall be accomplished which were written by the prophets concerning the Son of Man. For He shall be delivered to the Gentiles, and shall be mocked and scourged and spit upon : and after they have scourged Him, they will put Him to death, and the third day He shall rise again. And they understood none of these things, and this word was hid from them, and they understood not the things that were said. Now it came to pass, when He drew nigh to Jericho, that a certain blind man sat by the wayside, begging. And when he heard the multitude passing by, he asked what this meant. And they told him that Jesus of Nazareth was passing by. And he cried out, saying : Jesus, Son of David, have mercy on me. And they that went before rebuked him, that he should hold his peace. But he cried out much more : Son of David, have mercy on me. And Jesus, standing, commanded him to be brought unto Him. And when he was come near, He asked him, saying : What wilt thou that I do to thee ? But he said : Lord, that I may see.

And Jesus said to him : Receive thy sight : thy faith hath made thee whole. And immediately he saw, and followed Him, glorifying God. And all the people, when they saw it, gave praise to God."

Summary of the Morrow's Meditation.

We will meditate to-morrow upon the three days on which we are about to enter, and we shall see that we owe : 1st, to Jesus Christ ; 2d, to our neighbor ; 3d, to ourselves, to make them three days of penance and mortification. We will then make the resolution : 1st, to pass these three days in a spirit of recollection and of prayer, and to make during the time they last several fervent visits to the Blessed Sacrament ; 2d, not to yield to a worldly spirit during these three days, but, on the contrary, to practise on them some acts of penance and mortification. Our spiritual nosegay shall be the words of Our Lord to His apostles : "*Amen, amen, I say unto you, that you shall lament and weep, but the world shall rejoice, and you shall be sorrowful, but your sorrow shall be turned into joy*" (John xvi. 20).

Meditation for the Morning.

Let us adore Jesus Christ in the two facts which the gospel of the day places before us. On the one hand He predicts His Passion ; on the other He gives sight to a man who was born blind. The recital of these two facts is very striking in its actuality, in these days of license which show us on the one side the Passion of the Saviour renewed by the scenes of the carnival; on the other the world so blind to the things of God and of eternity. Let us make honorable amends to Jesus Christ for the licentiousness of the world, and do not let us allow Him to pass through these holy days without praying Him to enlighten and convert us. Let us fear, with St. Augustine, to permit Him to pass without our becoming better.

FIRST POINT.

We owe to Jesus Christ to make of these three Days three Days of Penance and Mortification.

We shall never be able to conceive all the sorrow which the licentiousness of the world occasioned the Heart of Jesus during these three days when, from the Garden of Olives, He saw them distinctly in the course of the centuries to come. It would be necessary, in order to conceive them, that we should love God like Him, understand like Him the enormity of sin, which despises the power of God, braves His justice, outrages His holiness, scorns His goodness, disowns His benefits: a horrible injury which He sees multiplied by millions of men during these three days; it would be necessary to love men as He does, understand, as He does, the misery of those souls who will not save themselves and are determined to be lost, treading His blood under their feet, rendering His sufferings useless, His love fruitless, in order to cast themselves head foremost into hell. O overwhelming agony! His soul is sorrowful even unto death (Matt. xxvi. 38). Now is it not the duty of friends to share in the sufferings of the friend whom they see is suffering, to console him and to visit him? Jesus Christ, exposed upon our altars, calls upon us to fulfil this great duty. We do not love Him if, neglecting to associate ourselves with His sorrows, we force Him to repeat the complaint which He breathed forth at an earlier period by the mouth of the prophet: "*I looked for one that would grieve together with Me, but there was none*" (Ps. lxviii. 21).

SECOND POINT.

We owe it to our Neighbor to make of these three Days three Days of Penance and Mortification.

Alas, the men who are ruining themselves are our brethren, and shall we not have pity upon them? (Matt. xviii. 33.) Do we love them if the misery into which they are about to cast themselves, says nothing to our heart, if we do not pray and perform penance for them? If it were only the loss of one single soul which was in question, he would have a heart of iron, says St. Augustine, a heart as hard as a diamond, if he could be insensible to it. What then ought we to feel, when we see so many who are ruining themselves? What ought we to feel, above all, in these days wherein a still greater number than usual enroll themselves beneath the banner of Satan? Oh, if we had true charity, if we loved our neighbor as ourselves, if we loved him as Jesus Christ loved us, according to the precept He has given us, what penances and mortifications should we not impose on ourselves during these three days for poor sinners! What are our dispositions, now that we are about to enter on these holy days?

THIRD POINT.

We owe it to ourselves to make of these three Days three Days of Penance and Mortification.

Our Lord, in fact, attaches to this practice a promise of salvation and a guarantee of predestination. O you, He says to His apostles, who remain faithful to Me in these days of tribulation and of trial, depriving yourselves of the pleasures of the world in order to bear in memory My cross, I promise to give you My kingdom, to make you enjoy the bliss of heaven, to establish you on thrones, where you will judge the twelve tribes of Israel (Luke xxii. 28–30). Elsewhere He promises those who are sad through love of Him, whilst the world rejoices, that their sorrow shall be changed into eternal joy (John xii. 20, 22). These are words which show us the portion of those who follow the world in these days of license, and the portion of those who follow the Lord. The one set of men pass their time in worldly amusements, the other in tears and penitential practices, but soon their tears shall be followed by a joy that never ends. In this alternative, which side will we take? Can we hesitate for one moment?

Resolutions and spiritual nosegay as above.

Monday after Quinquagesima.

Summary of the Morrow's Meditation.

In order to enter into the spirit of the Church during this season of adoration and expiation, we will meditate to-morrow : 1st, how the ingratitude of men makes the love of Our Lord show itself all the more in the Eucharist ; 2d, what are the duties incumbent on us from so much love that is disowned. We will then make the resolution : 1st, every day to make a visit of honorable reparation to the Blessed Sacrament for all the licentiousness of the world during these days of license ; 2d, to live in a more holy manner to-day, in a spirit of reparation for this licentiousness. Our spiritual nosegay shall be the words of David: "*I beheld the transgressors and I pined away*" (Ps. cxviii. 158).

Meditation for the Morning.

Let us transport ourselves in spirit before the altar where Jesus is exposed during these three days. Let us unite ourselves to the angels of the sanctuary who endeavor by their homage to make reparation for the transgressions of the earth; let us prostrate ourselves and be filled with sentiments of adoration, of love, of honorable reparation and of expiation.

FIRST POINT.

How greatly the Ingratitude of Men makes the Love of Our Lord shine forth all the more in the Eucharist.

When we are in the presence of the Blessed Sacrament we do not take sufficient account of what it cost Our Lord to descend so low. He was obliged to leave the bosom of His Father, where He was in God, where He was God, to come down to earth, and not to be received here, to take refuge in a stable and to escape death only by flight; He was obliged after thirty years of a laborious and hidden life, after three years employed in evangelizing the people and doing them good, and receiving in return calumnies, outrages, ignominy, and death, to survive by the institution of the Eucharist, in order to remain in the midst of the men who had treated Him so unworthily; and from that time, O miracle of love, what horrors has He not been obliged to submit to! The very first time that He celebrated the Eucharist He was buried alive, in company of the devil, in a stained conscience, in the soul of Judas. Since then, to what profanations, to what indignities has He not been subjected! The heretic denies Him, the bad Christian is wanting in respect, the impious man blasphemes Him; cupidity, for the sake of taking possession of the vessel wherein He reposes, casts Him on the ground and treads Him under foot. O Lord, how dearly Thy love costs Thee! Take Thy flight once more to heaven and free Thyself from so many outrages. —"No, I will not, I love mankind too much to separate Myself from them. Rather than deprive one single soul of the happiness of receiving and possessing Me, I will submit to all kinds of ingratitude and all kinds of profanations." And so He traverses the ages, always and everywhere disowned, except by a small number of souls who know how to appreciate His love. All along the road He is unworthily treated, sometimes left desolate in the solitude of the tabernacles, sometimes despised and insulted; finally He reaches our heart in Holy Communion, laden with eighteen centuries of profanations; it was the object to which all His desires tended, He has attained it; He is in us; He is ours; He is content. The more He has suffered on the road, the more love He shows us; it is all that He desired. O love, how incomprehensible thou art! O love! O love! can we ever appreciate thee as thou deservest? can we ever thank thee, ever love thee enough?

SECOND POINT.

Duties Incumbent on us in Consequence of so much Love that is Disowned.

1st. We ought to be profoundly touched by the outrages committed against so loving a God, above all by those which He receives during these very days of reparation and expiation. David and Jeremias are melted to tears, fall down fainting, and are parched with grief at the sight of the prevarications of the ancient people; how would they feel if they could but witness the far more guilty prevarications of the new people? St. Teresa could not think of them without giving vent to cries of sorrow and of desolation. Independently of irreverences and profanations, the simple fact of the manner in which the tabernacles were forsaken seemed to distract her. She gathered her nuns together and exclaimed to them: "My sisters, love is not loved, let us love the love which is not loved." All the saints have felt the same grief at seeing the love of Jesus in the Eucharist scorned by the ingratitude of men, and each time they had tidings of a profanation it was as though a sword had pierced their soul.

2d. We must make reparation for all these evils by fervent honorable amends to the Blessed Sacrament, offer it in expiation all our reverence, all the homage of the angels and the saints, all our actions and our life itself, protest that it would be a happiness to us to shed our blood to spare it the least offence or to repair it, and finally live to-day in a more holy manner than usual, visit the Blessed Sacrament with more love, communicate henceforth with more fervor, assist at the Holy Sacrifice more piously and more frequently.

Resolutions and spiritual nosegay as above.

Tuesday after Quinquagesima.

Summary of the Morrow's Meditation.

We will consider to-morrow how the mystery of Jesus outraged in the Eucharist causes to shine forth: 1st, His humility; 2d, His meekness; 3d, the perfection of His recollection. We will then make the resolution: 1st, to treat every one to-day with great meekness and humility; 2d, in the midst of the general license to maintain ourselves in a spirit of recollection and prayer. Our spiritual nosegay shall be the invitation of the Psalmist: "*Come, let us adore and fall down and weep before the Lord*" (Ps. xciv. 6).

Meditation for the Morning.

Let us adore Jesus, so humble and so meek, in the Blessed Sacrament, addressing to us, from the altar where He is exposed, His favorite maxim: "*I have given you an example, that as I have done to you, so you do also*" (John xiii. 15). Let us thank Him for these good words and for His holy example.

FIRST POINT.

How the Mystery of Jesus Outraged in the Eucharist causes His Humility to Shine Forth.

If there had been nothing else than this life hidden during eighteen centuries in the obscurity of the tabernacle, it would have been an act of wonderful humility. What must it not be to have to suffer the being forsaken by men, for the love of whom He is there! The majority abandon Him, some through neglect and indifference, others through contempt, and He spends weeks and months in solitude in this dark prison, submitting to the irreverence, the insults, the profanations, the sacrileges committed by many who come into the church, and that to such an extent as to be trodden under foot by malefactors, who steal away from Him the little vessel in which He reposes. O God of tabernacles, how humble Thou art! How, in presence of so much abasement, could I indulge in pride and self-love, in unreasonableness and susceptibilities? How could I desire to be preferred to others, to be brought into notice, and to be honored? Oh, rather would I say with David: "*I will make myself meaner than I have done, and I will be little in my own eyes*" (II. Kings vi. 32).

SECOND POINT.

How the Mystery of Jesus Outraged in the Eucharist makes His Meekness to Shine Forth.

We offend Jesus in the Blessed Sacrament by voluntary distractions, by irreverence, by giving license to our eyes and our tongues; by an irreligious deportment, by profanation and sacrilege; and yet amidst so many horrors He is meek and patient; He sees all, and appears as though He did not; He suffers all and is silent. During eighteen centuries He has not allowed us to perceive even one single time that He is displeased; not a movement of impatience, not a sign of ill-temper. He might launch His thunders against the profaners, open hell under their feet; but He loves better to say to us, "*Learn of Me because I am meek*" (Matt. xi. 29). What a marvel of meekness! And, also, what a lesson for me! What a condemnation of my hardness and of my impatience! I cannot bear to be opposed, that others should have their defects, and that they should not be angels! O Jesus so meek, teach me to suffer everything with meekness, without making any one suffer in any way, to moderate my quickness of temper, my anger, my bitter reproaches.

THIRD POINT.

How the Mystery of Jesus Outraged in the Eucharist makes His Love to Shine Forth.

Exterior things, above all those which hurt our self-love or which wound our feelings, preoccupy and distract us to such a point that, being entirely given up to outward things, we do not live either with God, in order to respect His presence and offer to Him our actions, or with ourselves, in order to study our defects, to follow the practices of virtue, and all the movements of our heart. Jesus outraged in the Blessed Sacrament teaches us quite the contrary. He does not allow Himself to be distracted by the contradictions of creatures, by contempt and outrages. Always recollected within, always calm, He prays in peace for the poor sinners who offend

Him, and the more they offend Him the more He prays for them, the more He recollects Himself in order to make honorable amends to the Divine Majesty for so many outrages. Is it thus that we keep our interior in a peaceful and recollected state amidst the tumult of exterior things, above all amidst events which wound our self-love?

Resolutions and spiritual nosegay as above.

Ash Wednesday.

Summary of the Morrow's Meditation.

We will consider to-morrow that the ceremony of the ashes invites us to sanctify Lent: 1st, by penance and mortification; 2d, by the thought of death. We will then make the resolution: 1st, cheerfully to embrace all the mortifications suitable to this holy season, fasting and abstinence, with all the crosses sent by Providence which we may encounter; 2d, to excite ourselves to do all these things well by means of the saying of St. Bernard: "*If you were destined to die to-day, would you do this or that?*"

Meditation for the Morning.

Let us adore the Spirit of God inspiring the Church to institute the ceremony of the ashes in order to teach us what are the pious dispositions in which we ought to pass the holy season of Lent. Let us thank Him for this excellent institution, and let us beg Him to enable us to understand it aright and to put it into practice.

FIRST POINT.

The Ceremony of the Ashes Preaches to us Penance and Mortification.

From the most ancient times ashes placed upon the head have been an emblem of penance and sorrow. Job, when he repented that he had pleaded the cause of his innocence in too unmeasured language, cries out: "*Therefore I reprehend myself and do penance in dust and ashes*" (Job xlii. 6). As a penance for the sacrilegious theft committed by Achan at the taking of Jericho, Josue and the ancients of Israel covered their heads with ashes (Jos. vii. 6). Later on Judith, Esther, Mardochai, Judas Machabeus employed these means for turning away the anger of heaven; Jeremias and all the prophets counselled this practice to the Jews who were stricken by God (Jer. xxv. 34). Lastly, Our Lord Himself speaks of ashes as the symbol of penance, when He says of the inhabitants of Tyre and Sidon that if they had seen the miracles which had been worked by Him in the midst of Judea, they would have done penance in sackcloth and ashes (Matt. xi. 21).

This it is which explains why the primitive Church distinguished by means of ashes penitents from the faithful; and on the first day of Lent she covered with ashes the heads of all her children without distinction, for this reason, says Tertullian, that every Christian is born to live in penance. The ceremony of the ashes is therefore as a seal which binds us to penance, in such a manner that to receive ashes on our head without having contrition in our heart is to simulate a feeling which we do not possess: it is hypocrisy. Let us enter heartily into the spirit of penance from the first day of this holy season. The interest of our salvation requires it; Jesus Christ explicitly declares so by His words: "*Except you do penance you shall all perish*" (Luke xiii. 5), and He teaches it still better by His example; the whole of His life was nothing more than a continual penance. All the saints, imitating Him, have performed penance, and what right have we to dispense ourselves from it? We have sinned many times, and all sin, even when remitted, demands penance. We have passions to conquer, temptations to combat, and penance is the surest preservative against both the one and the other. Let us here question our conscience: have we the spirit of penance suitable to the holy season of Lent?

SECOND POINT.

The Ceremony of the Ashes Recalls to us the Thought of Death.

"*O man,*" the Church says to us to-day, "*remember that thou art dust and unto dust thou shalt return.*" The Christian, therefore, who comes to listen to these words at the foot of the altar presents himself there as a victim who, submissive to the sentence pronounced upon him, comes to offer himself to be, when it shall so please the Sovereign Arbitrator of life and death, reduced to ashes and sacrificed to His glory. By this act he seems to say to God : Lord, I come to accomplish in spirit that which Thou wilt soon finish in deed. Thou hast resolved, in punishment of my sin, to reduce me on some coming day into ashes. I come myself to make the essay to-day. I forestall the decree of Thy justice, and I already execute it. The Church in making us begin the holy season of Lent by this solemn acceptation of death, by the great sacrifice of all that we have and of all that we are, gives us to understand that she looks upon the thought of death as the most suitable one to make us pass through Lent in a holy manner, that is to say, in flying from sin and in the practice of penance and of all the virtues. Who indeed can think seriously of death and not keep himself always in a state of readiness to appear before God, and not watch over his words and actions, and not mortify himself in order to expiate his past faults and satisfy divine justice, and not multiply his good works and increase his merits (Gal. vi. 10), and not detach himself from everything which will last for so short a time, and not repeat with St. Bernard, If I were to die after this confession, how should I perform it; after this communion, how should I dispose myself for it; after this conversation, how should I speak; at the end of this week, of this month, how should I spend the time? Let us beg of God to enable us rightly to understand this great lesson of death, and to make us deduce from it practical consequences suitable to the sanctification of Lent.

Resolutions and spiritual nosegay as above.

Thursday after Ash Wednesday.

Summary of the Morrow's Meditation.

We will meditate to-morrow: 1st, on the lesson of humility which the Church gives us in the ceremony of the ashes; 2d, the reasons why the Church gives us this lesson at the beginning of Lent. We will then make the resolution: 1st, to keep ourselves, during the whole of Lent, in a humble and contrite spirit, whilst beholding our nothingness and our sins; 2d, heartily to accept the penance of Lent, as being far less than we deserve. Our spiritual nosegay shall be the words of the Church: "*Dust thou art, and into dust shall thou return*" (Gen. iii. 19).

Meditation for the Morning.

Let us adore the Spirit of God inspiring the Church to institute the ceremony of the ashes, as being a powerful lesson of humility for all Christians. Let us thank Him for this holy inspiration, and let us ask of Him grace abundantly to profit by it.

FIRST POINT.

The Lesson of Humility which the Church gives us by the Ceremony of the Ashes.

If the Church places upon the head, which is the seat of pride, the ashes which are the symbol of the nothingness of human things, it is not only to preach to us thereby thoughts of penance and of death; it is also and above all for the purpose of saying to us: Do not inflate yourself so much, proud man. Remember that you are dust and ashes and that to dust you will return (Gen. iii. 19). Dust and ashes, behold whence thou comest; such is thy origin. God took a little clay, and of it formed the first man, whence have come all other men. Dust and ashes, behold what you are: a little clay formed into a man, says Tertullian. Now, is it suitable for clay to boast of what it is, to lift up itself through pride against Him who, animating it with His spirit, raised it through mercy above what it was? (Ecclus. x. 9.) Dust and ashes, behold what you will soon become, for you will return to dust (Gen. iii. 19). You will return thereto with the sensitiveness which takes offence, with thoughts of self-love and of complaisance in yourselves, with the desire to attract notice and be honored. All that at a certain day will disappear, and be nothing but a handful of ashes, will be lost in ashes, and vanish like ashes before the wind, after having been vile like it, sterile and useless like it. Even if you had equalled and surpassed in glory the most renowned men, in riches the most opulent of men, in enjoyments the men who have had the most enjoyments: all that, at the end, will be reduced to a few ashes; and these few ashes, it will not be possible to recognize; it will not be known to whom they belong; a blast of wind will disperse them in the air, and the very name of him from whom they come will be as entirely forgotten as though he had never existed. What a lesson of humility, well calculated to disabuse us of all the delusions of self-love, and to make us return to the humble sentiments we ought to have of ourselves! What folly to desire to be esteemed and honored, since we shall finally be reduced to a few ashes!

SECOND POINT.
Why the Church gives us this Lesson at the Beginning of Lent.

It is, 1st, because, without humility, all the mortifications of Lent would be devoid of merit. The Pharisees fast, said Jesus Christ in the gospel of yesterday; but as they do it in order to obtain the esteem of men, they do it without any merit, and receive their recompense upon earth. The reason is, because to esteem ourselves is to transgress against the truth, which tells us that we are nothing; and because to desire to be esteemed is to transgress against justice, which exclaims to us: To God alone be honor and glory (I. Tim. i. 17), to us, confusion (Baruch i. 15). Now, lies and injustice are incompatible with merit. It is, 2d, because without humility there is no true penance. True penance has for its basis the feeling of our misery, or the humiliation of the soul, which, confessing itself to be guilty, recognizes itself to be bound to make all sorts of reparations and satisfactions to divine justice. He who esteems himself may, like the Pharisee, perform exterior acts of penance, and say, like him: "I fast twice in the week; I pay the tax on all my goods;" but, at bottom, this penance cannot please Him who sounds hearts, and who takes delight only in truth. The Pharisee, notwithstanding his fasts, was none the less held in execration by God, for the sole reason that he esteemed himself and sought for the esteem and praise of others. Let us fear lest it may be so with us; and in order to prevent this misfortune, let us begin Lent in a spirit of humility.

Resolutions and spiritual nosegay as above.

Friday after Ash Wednesday.

Summary of the Morrow's Meditation.

In conformity with the spirit of the Church, we will meditate to-morrow on the crowning with thorns, and we shall admire in this mystery: 1st, a mystery of suffering and humiliation; 2d, a precious lesson for our salvation. We will then make the resolution: 1st, heartily to accept the mortifications and humiliations which may present themselves to us; 2d, often to make acts of contrition for our sensuality and self-love. Our spiritual nosegay shall be the words of St. Bernard: *"Let us be ashamed to be delicate members under a Head crowned with thorns"* (Serm. v., *in Fest. Omn. SS.*, no. 9).

Meditation for the Morning.

Let us, with very profound reverence, adore Jesus crowned with thorns, after having been cruelly scourged, and then presented to the Jews with a reed in His hand by way of sceptre, and an old garment of purple upon His shoulders by way of a royal mantle. O my Saviour! by all this they wish to turn Thy kingdom into ridicule; but under this insulting exterior I recognize Thee to be my Saviour and my God. I honor Thee, I praise Thee, I bless Thee under this unworthy disguise, which Thy love for me has caused Thee to accept.

FIRST POINT.

Jesus Crowned with Thorns a Mystery of Suffering and Humiliation.

1st. It is a mystery of suffering. For the thorns are strong and sharp; the soldiers drive them with heavy blows into the sacred head, which is the most sensitive part of the body, and they make the points pierce it so deeply that they make the little blood left by the scourging flow from it. From all parts the blood drops upon His adorable face, which is all disfigured by it. His holy humanity is thus wholly plunged in suffering; and the prophecy of Isaias is literally accomplished: *"From the sole of the foot unto the top of the head, there is no soundness therein; wounds and bruises and swelling sores; they are not bound up, nor dressed, nor fomented with oil"* (Is. i. 6). He accepts with calmness and resignation these dreadful sufferings, offering them to His Father for the salvation of the world. What heroic self-sacrifice! What incomprehensible love! O Jesus, how shall we ever realize such great charity!

2d. It is a mystery of humiliation. This great God is made a mock king and given up to public ridicule. A crown of thorns is placed on His head, to turn into ridicule the royal crown which He had a right to wear; He holds in His hand a reed as a sceptre; on His shoulders is a garment of purple instead of a royal robe; then they kneel down in His presence, and, mocking Him, they exclaim: "*Hail! King of the Jews*" (Matt. xxvii. 29). From ridicule they pass on to cruelty: "*Away with Him, away with Him! crucify Him!*" (John xix. 15) cries out the multitude. "*We will have no king but Cæsar*" (Ibid.). O my God, my true God! pardon these cries, pardon these sacrilegious mockeries! As for me, I will have no other king, no other God, save Thee (Ps. v. 3).

SECOND POINT.

Lessons to be Drawn from the Mystery of Jesus Crowned with Thorns.

1st. This mystery teaches us to weep over our past sins. On our knees, in presence of Jesus crowned with thorns, we ought to say to ourselves: Behold the work of my sins; behold what sufferings and what ignominies they have cost my God; and thinking thus, is it possible not to detest them, not to weep over them, not to wash them in our tears, mingled with the blood which flows from the adorable head of Jesus Christ? Is it possible not to join to grief for the past a firm resolution to lead a better and more Christian life for the future? 2d. This mystery preaches mortification to us; for, as St. Bernard says, it is a shame to be a delicate member of a Head crowned with thorns. It is a revolting contrast for the Saint of saints to be in suffering, and for me to be indulging in enjoyment; that Jesus should have His head crowned with thorns, and that I should seize every opportunity to procure myself pleasure when I can do so without committing a great crime. 3d. This mystery teaches us humility. For the crown of ignominy which Jesus wears is the condemnation of that crown of pride and ambition which is the object of our sweetest dreams. In choosing for His portion a crown of humiliation, Jesus willed to show us that He reproves the passion of wishing to make an appearance, to obtain notice and to rise above our fellows; how, on the contrary, He loves humble souls, who, content with God alone, do not seek to obtain favor in the eyes of the creature; who do good in secret, without noise, without thinking of renown, because virtue suffices them. Let us collect together these precious lessons in our heart, and let us conform to them our sentiments and our acts.

Resolutions and spiritual nosegay as above.

Saturday after Ash Wednesday.

Summary of the Morrow's Meditation.

We will meditate to-morrow: 1st, on the holiness of the season of Lent; 2d, on the means for sanctifying the season. We will then make the resolution: 1st, to guard our heart and our senses more carefully against sin and dissipation; 2d, to attach ourselves during this season to the reformation of the defect which it is the most important for us to correct. Our spiritual nosegay shall be the words of St. Paul: "*Behold now is the acceptable time, behold now is the day of salvation*" (II. Cor. vi. 2).

Meditation for the Morning.

Let us transport ourselves in spirit to the desert where Jesus spent forty days and forty nights. Let us contemplate Him, prostrate in presence of the majesty of God, His Father, kneeling down often with His face to the earth; pouring forth His soul, sometimes in adoration, praise, thanksgiving, sometimes in supplications to obtain from His Father mercy in favor of poor sinners; and joining with His prayers, uttered with tears in His eyes, an incomparable mortification, since during these forty days He neither eats nor drinks, nor has any other bed than the rocks and the bare ground; no other shelter than the vault of heaven. Let us render to Him, in this state, our homage of adoration, of admiration, of gratitude, and of love.

FIRST POINT.

Holiness of the Season of Lent.

First of all, Our Lord teaches it to us by His example. Although His life was always eminently holy, He gives to it, during these forty days, a special exterior character of holiness. 1st. He passes these forty days in retreat, in order to show us that we ought on our side to pass them in a holy recollection, that being the necessary condition to hearken to God in the bottom of our heart, to study Him and to know Him, to love Him and to enjoy Him; and at the same time in a spirit of reflection, which is a no less necessary condition whereby to know and reform ourselves. 2d. He spends it in prayer, to show us that we ought to be more faithful during this season to our exercises of piety, to pray more and with greater fervor. 3d. He subjects Himself during this time to the most severe mortification, in order to show us that we must, during Lent, grant less to sensuality, to tastes, to pleasures, to accept the privation imposed by the Church and to perform true penance. It is thus that Our Saviour, by His example, teaches us the holiness of the season of Lent; and this teaching of the Saviour is confirmed by that of the Church. For why these more frequent sermons, this multiplication of religious exercises; why the privations which are imposed upon us, if it be not to tell us that this season must be sanctified by penance? Oh, blessed be the Church for this teaching! In the course of our lives we so frequently forget penance; we are in such great need that we should be reminded of it every year; for penance is indispensable to us, whether it be for the expiation of our past faults or to prevent the backslidings which our weaknesses would infallibly cause us.

To all these lessons on the obligation of passing the holy season of Lent in a holy manner is to be added a powerful reason, drawn from the great mysteries of the Passion and of the resurrection of Our Lord, for which Lent serves as a preparation. For the fruit of these mysteries ought to be death to ourselves, and a new life all in God and all for God; now these mysteries produce fruit in us only in proportion to our having passed Lent in a really holy manner. We shall receive the plenitude of graces attached to their celebration if we reach the end of Lent in good dispositions;

but the contrary will take place if we incur the misfortune of passing such holy days in dissipation and want of reflection, in pusillanimity and tepidity. Let us rightly understand therefore the holiness of this season and the necessity of passing it in a better manner than the ordinary seasons of the year.

SECOND POINT.
Means for Sanctifying Lent.

1st. We must apply ourselves to the performing of our ordinary actions in a perfect manner, for therein lies all their holiness; that is to say, we must during this holy season pray and perform our spiritual exercises better, employ our time better, watch better over our words, give to each of our actions a higher degree of perfection and offer them to God in union with the penance of Jesus in the desert, in expiation for our sins and the sins of the whole world. 2d. We must be faithful to the fasting and abstinence prescribed by the Church, and if we cannot do it, or if we are dispensed from it, we must supply for it by interior mortification. Making our will fast by a spirit of obedience and condescension, our temper by always preserving meekness, our tongue by silence or discretion in our words, our mouth by the privation of certain indulgences which are not really necessary, our eyes by not letting them wander, our whole body by the modesty of its deportment and manners, lastly, our whole interior by retrenching useless thoughts, vain imaginations and the infinite desires which the heart allows itself to indulge in if we do not take care; these mortifications do no harm either to the head or to the breast, and they do great good to the soul. 3d. We must cheerfully accept the crosses that God sends us, such as infirmities, bearing with the tempers of others, their defects, and their wills which are contrary to ours. 4th. Lastly, we must fix upon a special defect, of which we must pursue the reformation during the whole of Lent. That, says St. Chrysostom, is the best of all fasts, because its effects are durable, not only during the whole year but throughout eternity. Are we resolved to embrace these different kinds of mortification? Let us have courage to do so.

Resolutions and spiritual nosegay as above.

First Sunday in Lent.

The Gospel according to St. Matthew, iv. 1-11.

"Then Jesus was led by the spirit into the desert, to be tempted by the devil. And when He had fasted forty days and forty nights, afterwards He was hungry. And the tempter coming said to Him: If Thou be the Son of God, command that these stones be made bread. Who answered and said: It is written, not in bread alone doth man live, but in every word that proceedeth from the mouth of God. Then the devil took Him up into the holy city, and set Him upon the pinnacle of the temple, and said to Him: If Thou be the Son of God, cast Thyself down, for it is written: That He hath given His angels charge over Thee, and in their hands shall they bear Thee up, lest, perhaps, Thou dash Thy foot against a stone. Jesus said to him: It is written again: Thou shalt not tempt the Lord Thy God. Again the devil took Him up into a very high mountain, and showed Him all the kingdoms of the world, and the glory of them, and said to Him: All these will I give Thee, if, falling down, Thou wilt adore me. Then Jesus saith to him: Begone, Satan; for it is written, The Lord thy God shalt thou adore, and Him only shalt thou serve. Then the devil left Him; and behold angels came and ministered to Him."

Summary of the Morrow's Meditation.

We shall see to-morrow in our meditation: 1st, that temptation, far from being an evil, may be turned to great advantage; 2d, on what conditions is the temptation thus changed into good. We will then make the resolution: 1st, to prevent temptations as much as possible by watchfulness over ourselves and union with God; 2d, promptly to turn away from the temptation as soon as we perceive it, and not allow it to trouble us. Our spiritual nosegay shall be the words of the apostle St. James: *"Blessed is the man that endureth temptation"* (James i. 12).

Meditation for the Morning.

Let us adore Jesus Christ tempted in the desert by the devil. It was, indeed, by far the greatest humiliation that a God could suffer, but He suffered it, because He knew that His example would encourage us in the midst of our trials and would teach us that the more a soul is dear to God, the more it ought to be proved by temptation (Tob. xii. 13). Let us thank Him for such great goodness.

FIRST POINT.

Temptation, far from being an Evil, may be turned to our Great Advantage.

No moral evil is possible excepting in so far as the will consents to it; as long as the door of the will is closed, the devil and the imagination may make a noise around the heart, but they cannot soil its purity. This is why Jesus Christ and all the saints have been subjected to the trials of temptation, without their trial having occasioned the least injury to their holiness. This is why to be cast down in temptation is unreasonable; it is either occasioned by self-love being annoyed at seeing itself so miserable, or else a want of confidence in God, who never fails any one who calls upon Him, or else it is the cowardice of a soul which imagines that it stands alone with its weakness, and without the help of God. Far from temptation being an evil, it may, on the contrary, be turned to our great advantage. For, 1st, it gives us an opportunity of glorifying God, since by generously resisting it we prove our fidelity to Him, we combat His enemies, and we triumph; 2d, it exercises us in humility, by revealing the evil basis which exists in us; in the spirit of prayer, by making us feel the need of having recourse to God; in vigilance, by warning us to mistrust our own strength and to fly from occasions of evil; in divine love, by its causing to shine forth the goodness of God, who is willing to lower His grace, to lower even Himself by communion to so depraved a level as ours; it prevents laxity, it awakens fervor, it gives to virtue a firmer and more solid character (II. Cor. xii. 9); it teaches us to know ourselves (Ecclus. xxxiv. 9). It gives the soul an opportunity to acquire more graces in this world, and more glory in the next world in proportion to the merits with which it enriches it, and renders it more worthy of God, like the saints of whom it is written, *" God hath tried them and found them worthy of Himself*" (Wis. iii. 5). This is why God said to the people of Israel: "*I would not destroy them from your face, that you may have enemies*" (Judges ii. 3), and Pope Leo also said in the same sense: "It is well for the soul to be afraid of falling, and to have a battle constantly to wage" (Serm. iii.). The faithful soul derives from temptation to evil the same fruit as from inspiration to good. It is an opportunity for it to tend towards perfection in the contrary virtue, with all the good-will of which it is capable. In temptations of the senses, it raises itself to the infinite glory of God, placed so high above all low and sensual views; in mental temptations it takes refuge in its nothingness; in temptations to pleasure it loves to embrace the cross. Is it thus that we profit by temptation?

SECOND POINT.

On what Conditions Temptations are Changed into Good.

There are certain conditions required before, during, and after temptation. 1st. Before temptation we must avoid all that exposes us to it or inclines us to evil, for example, dangerous society and books; looks which are kept too little in check; manners which are too free; the delights of an effeminate and sensual life. He who loves danger shall perish in it; he who counts upon his own strength shall be confounded. Mistrust is the mother of safety, and to expose one's self voluntarily to danger is to tempt God and to render ourselves unworthy of His help. On the other side, we must not be afraid of temptation, because by fearing it we give birth to it; the best is not to think of it and to be given up entirely to what we have to do. 2d. During temptation we must not amuse ourselves with it, under the pretext that it is slight, otherwise it will take the upper hand; but we must turn away promptly, firmly, and quietly from it; turn our back upon it with contempt, without even deigning to look at it; and if it produce some impressions upon us, we must disown them peacefully, applying ourselves wholly to present actions. Whoever fights with it risks soiling himself; and he who repels it with excessive efforts loses peace of heart, recollection of the mind, and the unction of piety. If we cannot succeed by these means, we must have recourse humbly to God, saying to Him: "O my Lord, how great is my misery; how wrong I should be still to cherish self-love, and how good Thou art to love a sinner such as I am! O Jesus! O Mary! O all ye angels and saints, bless the Lord who deigns to abase His love down to my nothingness." 3d. After the temptation we must forget it; reflection will bring it back to life. It is better to encourage ourselves peaceably to the repairing of the past wrong-doing, if any such has existed, by performing very perfectly the action in which we are engaged; by uniting ourselves to God, and casting ourselves into His arms with confidence and love, saying to Him as did the prodigal: "*Father, I have sinned against heaven and before Thee*" (Luke xv. 18), or like the publican: "*O God, have mercy on me, a sinner*" (Luke xviii. 13). Let us examine ourselves as to whether we have observed these rules, before, during, and after temptation.

Resolutions and spiritual nosegay as above.

Monday in the First Week.

Summary of the Morrow's Meditation.

We will meditate to-morrow upon the three temptations of Jesus in the desert, that is to say: 1st, an excessive care of the body and of health: 2d, the self-love which presumes upon itself and desires to attract notice; 3d, ambition and self-seeking. We will then make the resolution: 1st, to avoid excessive delicacy in the care of our body; 2d, to seek God only in all things. Our spiritual nosegay shall be the advice of the apostle St. James: "*Resist the devil and he will fly from you*" (James iv. 7).

Meditation for the Morning.

Let us adore Jesus Christ in the desert allowing Himself to be tempted by the devil, in order to teach us how to act in similar temptations. Let us bless this charitable High-Priest, who was willing to be tried by all kinds of temptations that He might resemble us in all things except sin, and let us put all our confidence in Him (Heb. iv. 15).

FIRST POINT.

First Temptation: an Excessive Care of the Body and of Health.

The devil draws near to Jesus, and says to Him, why dost Thou not eat? Thy body will not be able to bear it. Why dost Thou not tell these stones to change themselves into bread? (Matt. iv. 3.) Man does not live by bread alone, Jesus Christ answers, one word issuing from the mouth of God suffices to make him live. I have given to the Lord My life, My strength, My health; it is His property, He will take care of it; I abandon Myself to His providence. What a lesson for us is contained in these words of the Saviour, and He confirms them by His example. He lives during forty days in the desert, in a savage place, exposed upon a mountain to all the inclemencies of the weather; He fasts there during the whole of that time without tasting either bread or water; He watches during a great part of the night; and when He reposes, it is either upon a rock or upon the bare ground. He does not mean to tell us thereby to treat our own bodies with such severity, for it would try them too much. Health is a gift He has made us, and which He forbids us to deteriorate by excesses of any kind; but apart from this precaution, He interdicts us all sensual delicacies as regards our food, our clothing, our beds, and our lodging; He wishes that we should always feel that we are well placed where God wills we should be; and like St. Francis de Sales to go so far as to say, I am never better than when I am not well. He wills also that, following the example of St. Paul, we should not refuse to chastise our bodies and reduce them to servitude, whether for the purpose of expiating our past sins, or to prevent backsliding, or to appease the anger of God. Are these our dispositions?

SECOND POINT.

Second Temptation : Self-love Presumptuous and Jealous to Attract Notice.

The devil transports Jesus to the pinnacle of the temple, that He may be seen there by the whole world, and proposes to Him to cast Himself down from there, so that if He falls without injuring Himself He should be filled with a vain complaisance. Jesus Christ, repelling the temptation, renders Himself invisible to all the people, and returns by Himself to His beloved solitude. It is a beautiful example, which teaches us that, instead of wishing to attract observation and to show ourselves off in the presence of others, we ought: 1st, to show ourselves only through necessity, and always to tend to avoid esteem and praise, to be unknown and hidden; 2d, to keep ourselves on our guard against presumption, which esteems itself to be worthy of being honored, and thinks itself capable of bearing honors without ruining itself through pride. Let us here enter into ourselves and judge ourselves.

THIRD POINT.

Third Temptation: Ambition and Self-interest.

From the summit of a high mountain the devil places before the eyes of Jesus Christ all the kingdoms of the world, with their riches and their glory. I will give it all to Thee, he says, if Thou wilt prostrate Thyself before me and adore me. Get thee behind Me, tempter, answers Jesus Christ; it is written: Thou shalt adore the Lord thy God, and Him only shalt thou serve (Luke iv. 5). Thus ought every Christian soul to act. It ought to hold in horror all baseness, all intrigues, all insinuations tending to obtain the good graces of those who can obtain for it a good position, or enable it to rise to high places, or to maintain itself therein. It does not allow itself to be seduced by the bait of honors, and it cannot bend its knees to those who are the dispensers of them. It says, like the Apostle: "*To me it is a very small thing to be judged by you or by man's day; but neither do I judge my own self*" (I. Cor. iv. 3); in all things I think of nothing but my duty. If I please God, that suffices me, and all the rest is as nothing to me. O happy liberty! O holy freedom of a soul which has such dispositions! Let us examine ourselves before God as to whether these are our own dispositions.

Resolutions and spiritual nosegay as above.

Tuesday in the First Week.

Summary of the Morrow's Meditation.

As human weakness is so exposed to yield to the temptations which besiege it, we will meditate to-morrow on the Sacrament of Penance which Our Lord has established to raise us again after our falls, and we shall see: 1st, the excellence of the sacrament; 2d, the importance of receiving it properly. We will then make the resolution: 1st, often to thank Our Lord by pious aspirations for this admirable institution; 2d, to prepare ourselves better for our confessions. We will retain as our spiritual nosegay the very words of the institution of the Sacrament of Penance : "*Whose sins you shall forgive, they are forgiven them; and whose sins you shall retain, they are retained*" (John xx. 23).

Meditation for the Morning.

Let us adore Our Lord under the beautiful and amiable title of the Physician of our souls (Clem. Alex., *Pædag.*, lib. i. c. ix.). It is He who, by the Sacrament of Penance, cures all our evils, making of His own blood a salutary bath, wherein He washes away our stains and gives us back the beauty of our primary innocence. Oh, how greatly He merits our gratitude for so great a grace! "*What goodness to have made of His blood a remedy for our ills!*" (St. Augustine, *in Ps.* lviii.)

FIRST POINT.

The Excellence of the Sacrament of Penance.

1st. There is in this sacrament wonderful power. The Jews said : "*Who can forgive sins but God alone?*" (Luke v. 21.) And they were right, because God alone can dispose of His rights and remit the offence committed against Him. Yet, behold, by these words: "*I absolve thee,*" the priest exercises this superhuman power. By these few words he effaces all the sins of the soul, however enormous they may be; he chases away from it the devil; he reconciles it to God; he clothes it with the nuptial robe of charity; renders back to it the merits of its good works; re-establishes it in its rights to eternal life, and makes God enter once more into the heart whence sin had banished Him. God, having returned to the soul, fortifies it against backsliding, preserves it if it co-operates with grace, and often makes it enjoy delightful peace and consolation, so as to enable it to say: "*I shall be spotless with Him, and shall keep myself from my iniquity*" (Ps. xvii. 24). 2d. In the same proportion as there is power in this sacrament, there is also charity. Is it not indeed an ineffable miracle of charity that God, after having been offended by man, should have instituted in His Church a tribunal, not for the purpose of condemning and punishing, but for the purpose of pardoning; a wholly merciful tribunal, where no other accuser, no other witness is admitted save the guilty person himself; where repentance always obtains pardon, and a pardon accompanied by what we had lost through sin, the joy of a good conscience, rights to heaven reconquered, the title of a friend and servant of God? Is it not marvellous that Jesus Christ should have made of His precious blood a sacred bath, where the soul is purified and is clothed again with the beauty of innocence; an inexhaustible treasure of merits and graces which strengthen it in well-doing, dispose it to the practice of virtue, and assure it in heaven, if it perseveres, of a new weight of happiness and glory? Is it not marvellous, lastly, that confession, by the acts which accompany it, brings so much good to the soul? Self-examination teaches it to know itself; contrition makes it renounce its past faults; a firm resolution makes it enter upon a better way; and absolution gives it grace to walk in this new path. Let us here enter into ourselves and see if we have loved and esteemed the Sacrament of Penance as we ought to do, and whether we have not approached it with a feeling of constraint and of repugnance, of annoyance and sadness, or through routine and habit.

SECOND POINT.

The Importance of Receiving the Sacrament of Confession in a Proper Manner.

There is nothing more serious or more worthy of attention than the manner in which we confess; for it is a matter of life or of death, of heaven or of hell. A confession properly made is a source of grace; made through routine, without contrition for our faults, without a firm resolution to convert ourselves, is changed into sin, says St. Bernard. What a misfortune that the remedy for sin should itself become another sin, that we draw death from the very source of life, and that the blood of Jesus Christ falls upon us, as upon the Jews, for our loss and reprobation! Nevertheless, oh, how sad it is to think of! so great a misfortune is not as rare as we might believe. It is the misfortune of all who familiarize themselves with this great sacrament; who, losing sight of the lofty ideas faith gives us of it, confess hastily, from custom, and by way of acquitting themselves of a duty, without any serious examination of their consciences, without sorrow and a firm resolution, without any emotion of grace and of repentance for the effeminacy of their manners, their lukewarmness and pusillanimity; who are always being reconciled and are never penitent; who, always reciting the same formula, never weep over their sins; who, always confessing what most weighs upon their conscience, never confess their self-will, their bad temper, vanity, self-love, tendency to indolence and idleness, seeking after their own ease and indulgence in sensuality, lastly, all the passions, which are the principle of their faults. Let us examine ourselves as to whether we attach to the manner of confessing the great importance which this action merits.

Resolutions and spiritual nosegay as above.

Wednesday in the First Week.

Summary of the Morrow's Meditation.

As the first condition for confessing properly is thoroughly to examine our conscience, we will consecrate the following meditation to this examination. We will consider to-morrow: 1st, the importance of a daily examination of the conscience; 2d, the importance of the examination preparatory to confession. We will then make the resolution: 1st, every evening to perform with exactitude the examination of our conscience; 2d, to exercise special care in preparing ourselves well before confession. Our spiritual nosegay shall be the words of the Psalmist: *"I have thought on my ways, and turned my feet unto Thy testimonies"* (Ps. cxviii. 59).

Meditation for the Morning.

Let us adore Our Lord, who, in order to make us understand the importance of the examination of conscience, warns us by His saints that to perform it well is the sign of the elect, that to neglect it is the character of the reprobate (St. Greg., *Mor.*, ii., vi.). Let us thank Him for so useful a warning, and let us render to Him on that account all our homage.

FIRST POINT.

The Importance of the Daily Examination of our Conscience.

All the saints and all the masters of the spiritual life are unanimous in speaking of the daily examination of the conscience as the most efficacious means for correcting defects and advancing in virtue (St. Chrys., *in Ps.* iv.). The Pagan philosophers themselves enjoined their disciples to examine themselves every day on these three points: What have I done? How have I done it? What have I omitted to do? In point of fact, unless we make this examination every day we do not know ourselves. There exist within us vices so disguised, ill regulated feelings so hidden, passions so subtle, that we do not perceive them except by means of serious reflection. It is with the soul which does not examine itself, or else examines itself badly, as it is with a vineyard left fallow, which, for want of being cultivated, is covered with thorns and briars; or it is like the man of business who, for want of rendering to himself every day an account of his position, allows the state of his fortune to get worse without having the least idea of it. For want of examination vices increase in the soul and virtues disappear from it; without our remarking it, the state of the conscience becomes constantly worse and worse, and such is the ignorance we are in of ourselves, that we do not even suspect it. The soul languishes, loses its strength, is no longer on its guard against temptations and dangerous occasions, and in this state it is on the point of being lost. By a daily examination, on the contrary, we remark our failings and we repair them; we say to ourselves every evening: "I have committed such and such a fault to-day, I will correct myself to-morrow; I will observe such or such a bad inclination in my heart, I will fight against it." Every day we say to ourselves: "I shall have to render an account this evening of the employment of my time, of my fidelity to grace," and this thought will awaken vigilance, excite my attention, and hinder bad habits from taking root. Moreover, the sight of our wretchedness, which daily examination keeps always before our eyes, preserves humility, makes presumption keep aloof, disposes us to make a good confession by means of a clearer knowledge of our faults. Lastly, daily examination, when it is accompanied by perfect contrition, as it ought always to be, protects the soul from the danger of sudden and unprovided death, since contrition stands in place of the sacrament when we cannot receive it. Let us examine ourselves as to whether we attach to this exercise all the importance it deserves, and whether we make it every day at a regular hour.

SECOND POINT.

The Importance of the Examination of Conscience before Confession.

We have to do here with a holy confession or else a sacrilegious one. If by a notable fault in our examination we omit to accuse ourselves of a single mortal sin, the confession is null and the absolution a sacrilege: what, then, can be more serious than this? If, on the contrary, each time that we confess our examination is made as it ought to be, confession purifies the soul from the past and renders it strong for the future: what can be more consoling? Nevertheless, how many times does it not happen that we make our confessions lightly, and content ourselves with casting a rapid glance over the time which has elapsed since our last confession? It is a very serious matter; it has to do with our eternity.

Resolutions and spiritual nosegay as above.

Thursday in the First Week.

Summary of the Morrow's Meditation.

To-morrow we will meditate upon the manner of making the examination of conscience, and we shall see: 1st, the character of this examination; 2d, the acts which ought to accompany it. We will then make the resolution: 1st, to observe in our examination the rules laid down by the saints; 2d, to bring to it above all a sincere regret for our sins and a firm resolve to correct them. Our spiritual nosegay shall be the words of King Ezechias: "*I will recount to Thee all my years in the bitterness of my soul*" (Is. xxxviii. 15).

Meditation for the Morning.

Let us adore in Jesus Christ the perfect knowledge which He has of our sins. Not a single one escapes Him; He knows all their circumstances; He penetrates into all the malice of them—very different in this from men, who often perceive only the outward appearances, and who permit themselves to be deceived by the prejudices and the disguises of self-love. Let us bless this amiable Saviour, who is so willing to make us participate in His divine light that we may thoroughly know all our sins.

FIRST POINT.
Characteristics of the Examination of Conscience.

This examination ought to be made with exactness, severity, and calmness.

1st. *With exactness;* that is to say, that it ought to embrace, 1st, the evil we have committed, the good we ought to do and which we have not done, and even the good we have done badly; 2d, sins against God, against our neighbor, against ourselves; external sins, coming from the senses, especially of the tongue; internal sins, which are thoughts, desires, attachments which have not God for their object; 3d, the number of times that we have fallen, the principle or the source of our faults, their circumstances and their consequences. In order to attain to this exactness it is easy to conceive that we must bring to it an attentive search, not stopping at the surface, but penetrating to the bottom of things. Is it thus that we make it?

2d. *With severity;* that is to say, without listening to self-love or natural tenderness, which leads us to excuse ourselves, to hide our faults from our own eyes, or at least to lessen them; we must examine ourselves as a judge would examine a criminal, or as though we ourselves were examining a stranger. A too indulgent examination sees only trifles, where serious faults really exist; for example, certain calumnies, aversions, or jealousies; in a certain kind of luxurious expenditure, a certain kind of loss of time, certain kinds of vanity, and desires to attract notice. Do we not often delude ourselves upon many of these points for want of bringing enough severity to our examinations of conscience?

3d. *With calmness;* that is to say, we must not torture the conscience with the fear of forgetting certain faults, but make the examination with the tranquillity of the accountant who is making up his accounts; of the judge who is summing up a lawsuit; of the doctor who studies a malady. Why should we trouble and distress ourselves? A defect of memory is not imputed to us as a sin. He who has an upright intention to say everything, a sincere desire to make himself known, a frank will to dissimulate nothing, and who employs a reasonable time in his examination, does all that is necessary. God does not ask that we should tell Him all we have done, but only all that we remember, and anything that we forget is remitted as though we had accused ourselves of it. Consoling thought, well fitted to make us perform our examinations with calmness, simplicity, and freedom of heart.

SECOND POINT.

Acts which ought to accompany the Examination of Conscience.

This examination would be of little use to us if it were nothing more than a philosophical study of the state of our conscience. In order to be really useful to us it ought to be accompanied by the three principal exercises of piety: 1st, before the examination we must place ourselves in the presence of God; adore Him as our judge; keep ourselves humbly at His feet like poor criminals, and beg Him to give us His light, which alone can discover to us our faults without awaking our passions ; 2d, after the examination we must excite ourselves to repentance for our faults with sighs and tears ; make strong resolutions to correct them, and fix upon what we will do with that end in view : vague and general resolutions end in nothing ; 3d, we must place ourselves in the state in which we would desire to be at the hour of death, and by uniting ourselves to the heart of Jesus Christ ; to that heart so full of horror for sin, and of love of penance, which is the expiation of it. Is it thus we make our examinations ? It is for want of being faithful to these holy practices that our numerous examinations of conscience have not changed us. We have condemned sin without condemning the sinner, and we have remained always the same.

Resolutions and spiritual nosegay as above.

Friday in the First Week.

Summary of the Morrow's Meditation.

In conformity with the Roman liturgy, we will meditate to-morrow : 1st, on the nails that fastened Jesus to the cross ; 2d, on the lance which opened His sacred side. We will then make the resolution : 1st, to make frequent acts of love to the crucified Jesus, and not to refuse Him any sacrifice ; 2d, to excite ourselves to this love by often kissing the feet, the hands, and the sacred side of our crucifix, which recall to us the wounds made by the nails and the lance in the body of the Saviour. Our spiritual nosegay shall be the words of St. Paul: "*The charity of Christ presseth us*" (II. Cor. v. 14).

Meditation for the Morning.

Let us transport ourselves in spirit to Calvary; let us there contemplate Jesus on the cross; let us press our lips to His feet and His hands pierced by the nails, and to His sacred side opened by the lance; let us mingle our tears with the blood which flows; let us love the God who has so loved us.

FIRST POINT.

On Devotion to the Nails which attached Jesus to the Cross.

Doubtless if we see in these nails only a piece of common iron, they do not deserve any worship; but if we look upon them as empurpled with the divine blood which they caused to spring forth from the veins of Jesus; as impregnated with His flesh which they tore; as consecrated by their sojourn in that same flesh, who is there that does not see how venerable they are and what lessons they impart to us? 1st. They recall that spirit of obedience and submission which is the real spirit of Christianity, and which is so opposed to the spirit of the world which dreams of nothing but liberty and independence. The executioners said to Jesus: Stretch out Thy hands, stretch down Thy feet, that we may pierce them with nails. Jesus obeys; they nail Him to the cross, and He loses all power of movement; 2d, in fastening Jesus by a visible link to the cross, these nails make us to feel more acutely the invisible links of His charity which kept Him so firmly attached to it; 3d, they tell us how we ought to weep over the bad use we have made of our hands and our feet, the irregularity which marks our acts and our affections, since it cost Jesus so much to expiate them; 4th, they preach patience to us; for who can conceive all that Jesus suffered, and the patience with which He suffered it, whether when the executioners, driving the nails with great blows of their hammers into the most sensitive and nervous parts of the body, made four great wounds in it, whence flowed forth four rivulets of blood; or when, after having raised the cross, they let it fall into the excavation below with a terrible concussion which renewed all His sufferings and enlarged His wounds? O my Saviour, I adore Thee raised between heaven and earth, as the victim upon the altar of sacrifice to reconcile the one with the other; as our doctor and our master teaching us all truth. I love, O Jesus, those outstretched arms which tell us that Thou dost embrace us in Thy love; that head bristling with thorns, which, having nothing on which to lean, bows itself down to give us the kiss of peace and of reconciliation ; that breast bruised with blows, but which still rises and falls with the beatings of love which agitate Thy heart; those hands which the weight of the body suspended in the air drags violently down ; and those feet, the wounds in which widen under the weight of the body with which they are laden ! Oh, who would not love Him whose nails reveal to us so much love !

SECOND POINT.

On Devotion to the Lance which Opened the Sacred Side of Jesus.

St. Bonaventura had a very special devotion for the lance which opened the sacred side. O happy lance ! he exclaimed, which wast worthy to make that opening. Oh, if I had been in the place of that lance, I would not have come out again from the side of Jesus ; I should have said : Here is the place of repose which my heart has chosen, I will dwell therein forever, and nothing shall ever be able to tear me away from it. At least, adds the pious doctor, I will keep myself near the opening ; there I will speak to the heart of my master and I shall obtain all that desire. St. Bernard thought the same. That most blessed lance, he said, although handled by a soldier, was guided by Jesus, who thus opened to us His sacred side in order to show us thereby His divine heart all palpitating with love for us, or rather that He might give it us, and make us enter into it. O mysterious entrance! it is by it that we reach the heart of Jesus; that heart so good, so amiable, so loving and entirely ours; that heart, the true holy of holies, where the soul, shutting up itself therein, prays, adores, and loves as it ought; the true ark of salvation in which all should take refuge who do not desire to perish in the deluge of the world. Oh, a thousand times venerable, a thousand times blessed be the lance, which opened to us the door by which have come to us such great blessings, so many graces, and so much love!

Resolutions and spiritual nosegay as above.

Saturday in the First Week.

Summary of the Morrow's Meditation.

We will meditate to-morrow, 1st, upon the nature and importance of the particular examination; 2d, upon the manner of making it. We will then make the resolution: 1st, to be henceforth very faithful to this exercise; 2d, to make it according to the rules laid down by the teachers of the spiritual life. We will retain as our nosegay the words of Jeremias: "*I have set thee this day to destroy and to build and to plant*" (Jer. i. 10).

Meditation for the Morning.

Let us adore Our Lord who, animated by a desire to render us perfect, teaches us, by means of masters of the spiritual life, the exercise of the particular examination as one of the most powerful means of salvation. Let us thank Him for this goodness, always attentive to what may be useful to the soul.

FIRST POINT.

The Nature and Importance of the Particular Examination.

There is this difference between the general and the particular examination, that the first embraces the whole of the sins which we have committed during the day, or the space of time to which the examination is limited, whilst the particular examination has for its aim a special subject, for example, a vice, a virtue, an exercise, above all, the besetting sin, which is the weak side by which we are most exposed to lose our souls. This exercise is of great importance: 1st, because it is right first of all to provide for the safety of the part where our soul is most exposed to peril; now each man has in his soul a feeble side by which the devil principally attacks him, imitating therein the general who, in order to take a town, studies the weakest point and directs all his efforts to it; 2d, because our attention, disseminated over all our miseries at one and the same time, acts less efficaciously than when it concentrates its energies on one particular point; 3d, because the principal vice having been overcome, we shall easily conquer the others, in the same manner as an army which has lost its chief is easily put to rout. Let us here examine our conscience. Have we esteemed, as we ought to do, the particular examination, and do we make it assiduously every day? Do we bring to it all the attention necessary to search out and become acquainted with our least faults in regard to the matter which is the object of it? Do we not sometimes perform it with a great deal of negligence, because we do not appreciate all the importance of it? Do we not imagine that a minute research into our smallest failings would render us scrupulous, and that we can dispense ourselves from it?

SECOND POINT.

The Manner of Making the Particular Examination.

In order to do this well, we must: 1st, lay down the subject clearly, choosing the vice, the passion which is the most ordinary source of our temptations and our failings, or the virtue which is the most opposed to the vice; for example, humility for the proud, fraternal charity for those who are the most exposed to fail in it, mortification for too effeminate souls, gentleness and patience for the ill-tempered, chastity for tempted souls, conformity to the will of God, the perfection of our ordinary actions, and other practices, according to the need of each person. Let us here examine ourselves. Have we a subject of particular examination which is well adapted to the needs of our souls? If we have not, let us fix upon one from to-day. 2d. The subject once chosen, we must divide the parts and the relations of it, examine ourselves during a certain time, for example, on words contrary to humility, charity, or patience; later on, upon the acts which are contrary to these vices; later on still, upon the thoughts and sentiments contrary to them. 3d. After having examined ourselves, we must set down in writing, or at least retain thoroughly in our memory, the number of failings, and impose upon ourselves a penance proportioned to the number of falls, for example, a trifling alms put in reserve somewhere; it will be, in addition to the good work, an easy means for becoming acquainted with our failings. 4th. This examination being thus made beneath the eyes of God, in presence of Jesus Christ our judge, we must disavow our faults, ask pardon for them, make resolutions to preserve ourselves better from them in future, and pray to obtain the grace of our conversion. Is it thus that we daily make our particular examination?

Resolutions and spiritual nosegay as above.

Second Sunday in Lent.

The Gospel according to St. Matthew, xvii. 1–9.

"At that time Jesus taketh unto Him Peter, and James, and John his brother, and bringeth them up into a high mountain apart ; and He was transfigured before them. And His face did shine as the sun, and His garments became white as snow. And behold there appeared to them Moses and Elias talking with Him. And Peter answering said to Jesus: Lord, it is good for us to be here ; if Thou wilt, let us make here three tabernacles, one for Thee, and one for Moses, and one for Elias. And as he was yet speaking, behold a bright cloud overshaded them ; and lo, a voice out of the cloud saying, This is My beloved Son in whom I am well pleased ; hear ye Him. And the disciples hearing fell upon their faces, and were very much afraid ; and Jesus came and touched them, and said to them, Arise, and fear not. And they lifting up their eyes saw no one, but only Jesus. And as they came down from the mountain, Jesus charged them saying: "Tell the vision to no man till the Son of man be risen from the dead."

Summary of the Morrow's Meditation.

We will consecrate the whole of next week to meditating upon the gospel of to-morrow which contains the recital of the mystery of the Transfiguration. We will meditate to-morrow upon the two first circumstances which are the choice made by Jesus Christ for the Transfiguration: 1st, of a place apart and solitary (Matt. xvii. 1); 2d, of a high mountain. We will then make the resolution: 1st, not to frequent the world excepting from necessity and to love to be alone with God only; 2d, to detach ourselves from everything to which our heart is still bound here below. Our spiritual nosegay shall be the words of our meditation: "*Jesus led His disciples to a high mountain apart.*"

Meditation for the Morning.

Let us transport ourselves in spirit upon Thabor; let us admire the choice which Our Lord made of this solitary place, apart from the world, of this high mountain which rises nearer heaven. There is in this choice a double secret reason. Let us beg God to enable us to understand it.

FIRST POINT.

Why Our Lord Chose for His Transfiguration a Place apart from the World.

By this choice Our Saviour wills to teach us that it is not in the midst of the world and of worldly thoughts that God reveals Himself to the soul and makes it pass from the miseries of the old man to the splendors and virtues of the new man. In order to see God, to hear Him, to enjoy Him, and to be transformed into Him by His grace, the first condition required is interior solitude; that is to say, the calm of the soul closed to the turmoil of creatures, open to God alone and to His divine inspirations, the peace of recollection beneath the eye of God. As long as we indulge in dissipation of the mind, the wanderings of the imagination, the affairs of this world, wooed by attachments, the tumult of useless thoughts; as long, finally, as we do not live in retirement in the solitude of the heart, God will not show Himself to us, and He will only be to us the unknown God of Athens. His amiabilities and His infinite perfections will not touch us; we shall not love Him, and we shall have no desire to love Him. Strangers to God, we shall be no less strangers to ourselves; we shall not know ourselves, and we shall see nothing in ourselves to correct, nothing to reform, no reason to humble ourselves, to mortify ourselves, to renounce ourselves ; and the whole of our life will be spent in forgetfulness of God and in ignorance of ourselves. O dissipation, what harm you do to the soul ! O holy recollection, how necessary you are to it ! Lead me, O Lord, like Thy apostles, into solitude, and keep my mind and my heart always shut up there !

SECOND POINT.

Why Our Lord Chose a High Mountain for His Transfiguration.

This elevated place, where the apostles were raised above the objects amidst which they had hitherto lived, signifies that, in order to enjoy God, to merit His grace and to sanctify ourselves, we must have a heart raised above all sensible things ; a heart greater and higher than the world ; we must tread under foot everything that formerly attached us. As long as we cling to anything here below, as long as there is anything on earth which holds us in chains, we shall only crawl miserably in the same paths, and turn round and round in the labyrinth of our miseries, instead of advancing in virtue and strengthening ourselves. If our soul had the wings of the dove, for which the prophet king prayed, in order to fly away into the bosom of God, so long as it remains attached even by only one single thread, it will never do anything else but struggle and torment itself painfully about what retains it, without ever being able to take its flight. But also, if the soul has at last courage enough to break its chains, if it allows itself to be led by Our Lord up the mountain, and if from thence it treads under foot all the vain objects of its attachments, its progress in perfection will immediately begin. In a single day, and with less trouble, it will make more way than it did during the whole time that it dragged along after it the weight to which it was attached. Nothing will retard it on its course, nothing will restrain or distract it from its progress; it will advance easily and freely; for as the Imitation says: *"What is more free than he who is attached to nothing upon earth?"* If, then, we desire to become solidly virtuous, we must detach ourselves from all that flatters vanity, from all that nourishes effeminacy, from all which piques curiosity, from the frivolities which amuse, the affairs which distract, the society which dissipates us; we must renounce the passion of pleasure and enjoyment, we must not any longer cling so much to earthly comforts; we must satisfy our necessities with discernment, take things only in so far as we really need them, and touch them as it were only lightly and for the moment, like the soldiers of Gideon, or like Jonathan, who took the honey on the end of his staff without making a halt; above all, we must be detached from ourselves, from our tastes and our likings, from our self-will and its caprices, from our self-love and its ambition, which seeks to place itself in all that it says, and to find itself in all that it does; we must break off the excessive care for our health which renders us sensitive in respect to all that annoys and restrains the senses; we must, lastly, rise above ourselves (Lam. iii. 28), and under the penalty of being lost, we must empty our heart of all that is not God. At what point have we arrived in regard to this universal detachment? It is a more serious question than we think. Let us think of it seriously, and let us labor at it every day.

Resolutions and spiritual nosegay as above.

Monday in the Second Week.

Summary of the Morrow's Meditation.

As the gospel, by representing Our Lord transfigured whilst He was at prayer, reveals to us thereby that prayer is the means whereby to bring down on us all the graces of Heaven, we will meditate to-morrow: 1st, on the necessity of prayer; 2d, on the conditions for performing it well. We will then make the resolution: 1st, to be very exact in preparing our subject for meditation, both in the evening and the morning, and always to begin the day with this exercise; 2d, to maintain within ourselves during the day the good thoughts and the good sentiments of the morning's meditation. Our spiritual nosegay shall be the words of the Gospel: "*Whilst He prayed the shape of His countenance was altered*" (Luke ix. 29).

Meditation for the Morning.

Let us adore Jesus Christ at prayer upon Thabor. What a recollected and fervent prayer! What a beautiful spectacle in the light of heaven and of earth! Whilst He was absorbed in God, His face became as shining as the sun, and His garments as white as snow. Let us thank our divine Saviour for having revealed to us the value of prayer, to give us a love for it and to teach us to practise it.

FIRST POINT.

The Necessity of Prayer.

All the saints are unanimous in telling us that meditation is essential to salvation (Luke xi. 1); that a day without prayer is a lost day; that without prayer faith languishes, together with the appreciation and the sentiment of Christian truths and of our divine mysteries. He who does not meditate upon God and upon His infinite loving kindnesses is cold and indifferent towards

Him; he who does not reflect upon his duties does not any longer feel the importance of them: he neglects them, or accomplishes them badly. Without meditation no prayer can be well performed. It is impossible, says St. Teresa, to recite even the *Pater* properly; habit, routine, and wandering thoughts reduce the prayer to a simple movement of the lips, in which the heart takes no part. "*My heart is withered,*" says David, "*because I forgot to eat my bread*" (Ps. ci. 5). Without meditation there is no longer any recollection, any humility, any love, any virtue, says St. Bonaventura. Lastly, like to the soldier without any weapons, exposed to all the assaults of the enemy, we are defenceless against the devil, against the world, against our own heart. With prayer, on the contrary, faith becomes every day more lively; we appreciate God and heavenly things, as well as the nothingness of the world and the greatness of eternal graces; we see our faults and our defects, together with the remedies necessary to apply to them; the fire of the passions is extinguished and gives place to holy love. It is in meditation that the sacred fire is kindled (Ps. xxxviii. 4), and the whole life is changed and renewed (Luke ix. 29). Formerly we were frivolous; we did not reflect; we were pusillanimous and devoid of energy, irascible, attached to ourselves and our own senses; by means of meditation we become serious, recollected, courageous and fervent, gentle and modest, humble and without pretension. It was all this which inspired St. Augustine with those beautiful words: "*To know how to pray well is to know how to live well.*" Is this the idea we have of prayer?

SECOND POINT.
Conditions Requisite for the Right Performance of Prayer.

There are three principal ones: the habit of recollection, the detachment of the heart, the pacification of the passions. 1st. The habit of recollection. The mind which is essentially dissipated is incapable of prayer; accustomed to fix itself upon nothing, to fly without ceasing from one object to another, it follows its customary habit in meditation. In vain God speaks to it—it does not listen to Him; or, if it listen to Him, it does not reflect upon what He says, and lets its thoughts wander to other things. It is only in the silence of the recollected soul that God speaks, that His voice is heard, that the soul meditates upon it, that it appreciates it and profits by it (I. Imit. xx. 6). 2d. Detachment of the heart is necessary. The heart which has attachments takes possession of the spirit, carries it away captive, and tyrannizes over it. It desires to reflect upon God and upon its salvation, but attachments to other things pre-occupy and absorb the mind.

It cannot any longer occupy itself except with the things to which it clings; it is a cloud which prevents the light of God from being seen; it is a noise which hinders from hearing His voice. It desires to rise towards heaven, but the soul is attached to the earth; it may indeed agitate and torment itself, but it is impossible for it to take its flight. The soul which is detached, on the contrary, free from all ties, rises easily to God, converses with Him, and remains united to Him. 3d. The pacification of the passions is necessary. As long as the heart is possessed by any passion which it will not renounce, it will be ill at ease, troubled, and incapable of fixing itself upon God. It will be like the sick man devoured by fever, and, unable to sleep, he turns ceaselessly from side to side, and to be obliged to remain in the same position is unbearable to him. He, then, who desires to succeed in meditation and to make progress in it, must labor every day to master his passions until he has extinguished them. It will be only then that he will enter into that tranquillity of the soul which allows of sustained reflection and of an enduring union with God. Let us examine whether we have fulfilled the three conditions necessary for succeeding in meditation.

Resolutions and spiritual nosegay as above.

Tuesday in the Second Week.

Summary of the Morrow's Meditation.

The mystery of the Transfiguration upon which we are meditating this week makes three marvellously beautiful truths to shine forth: 1st, the greatness of Jesus Christ; 2d, the power of His mediation; 3d, the authority of His teaching. After considering these things we will make the resolution: 1st, to entertain within ourselves a great reverence for Jesus Christ, and a great confidence in His mediation; 2d, to imitate Jesus Christ and obey His inspiration. Our spiritual nosegay shall be the words of the Gospel: "*This is My beloved Son, in whom I am well pleased; hear ye Him*" (Matt. xvii. 5).

Meditation for the Morning.

Let us transport ourselves in spirit upon Thabor, and let us listen with great devotion to the panegyrics which God the Father pronounces upon His Son. Let us love the Father who thus praises, and the Son who is thus praised.

FIRST POINT.

The Greatness of Christ Revealed upon Thabor.

If we have preached to you, says St. Peter to

the faithful in his second epistle, the power and the coming of Christ, it is that "*we have not followed cunningly-devised fables when we made known to you the power and presence of Our Lord Jesus Christ: but having been made eye-witnesses of His majesty. For He received from God the Father honor and glory, this voice coming down to Him from the excellent glory : This is My beloved Son, in whom I have pleased Myself; hear ye Him*" (II. Pet. i. 16, 17). Oh, how great He is, He whom we adore in our tabernacles! And with what holy trembling, what profound religion, ought we to appear in presence of His infinite grandeur, not less real when He veils it through love and humility beneath the eucharistic species than when He reveals it on Thabor before the dazzled eyes of His apostles! It is the Son of God, not by adoption, by resemblance, by elevation, like the just, but by nature, by identity of essence, equal in all things to His Father; like Him, all-powerful, eternal, immense, infinite in all perfection, holy of holies, God of the universe, Creator of all things. Let us prostrate ourselves before so much greatness, and ask pardon of Him for having been so often wanting in respect in church, in prayer, in the habitual disposition of our heart.

SECOND POINT.

The Power of Christ's Mediation Revealed on Thabor.

Jesus Christ had already declared Himself to be our mediator with His Father by those sweet words which He said to His apostles: Ask My Father in My name; but upon Thabor God the Father reveals to us the power of this mediation by proclaiming Him to be His only Son, His well beloved, the object of all His complaisance, and, consequently, not only possessed of all power over His heart, but the only one by whom all prayers must be presented; the only one who infallibly obtains a gracious answer to them. It was the design of the Father that we should be redeemed and sanctified by this well beloved Son; it is also His design that all our prayers should be offered through Him and always answered by Him, on account of the great reverence He bears Him (Heb. v. 7). What a consolation for us to have such mediation! With what fulness of confidence we ought to address all our prayers to Heaven through Him! Do we not often forget these means of assuring the effect of our prayers?

THIRD POINT.

The Authority of Christ's Teachings Revealed on Thabor.

It ought to be an immense consolation for us to be the disciples of a Master and Doctor respecting whom Heaven proclaims the divine mission in a manner so lofty and solemn. *Hear ye Him,* said the heavenly voice ; listen to His teachings, not only because they reveal to you the dogmas of faith, which it is your duty to believe, spite of what the senses and reason seem to you to contradict ; but also when He preaches moral and practical truths to you, telling you that the happiness of this life consists in poverty, in contempt, and in suffering, that we must renounce and hate ourselves, oppose and do violence to ourselves, deprive and crucify ourselves without pity. *Hear ye Him* when He teaches you by the language of His examples (H. Pet. i. 17). He always led a laborious and hidden life, all His days were passed in suffering ; He placed Himself below all others, even at the feet of His disciples ; He was meek and humble of heart, accepting as His portion poverty, opprobrium, humiliation, and suffering. *Hear ye Him* when He speaks to you by the secret voice of His inspiration. His grace is always at the door of your heart, urging you to lead a better life, and to put aside the life which is only according to nature, entirely human, the life of frivolity and dissipation, the life of routine and habit, the life which is eternally the same, void of any reformation of defects as well as of any progress in virtue. Submit yourselves, in a word, to the expostulations of the grace which urges you. Happy he who listens to it in the peace and silence of the soul, and who, after having listened, generously obeys it (III. Imit. i. 1). Is it thus that we act?

Resolutions and spiritual nosegay as above.

Wednesday in the Second Week.

Summary of the Morrow's Meditation.

We will consider to-morrow in our meditation: 1st, that the mystery of the Transfiguration ought to kindle in us holy desires for heaven; 2d, that these holy desires are very useful to the soul. We will then make the resolution: 1st, to detach ourselves from the earth and no longer to love anything excepting heavenly things; 2d, often to give vent to holy desires in the form of ejaculatory prayers. Our spiritual nosegay shall be the words of St. Bernard: "*How beautiful thou art, O my country, how beautiful!*"

Meditation for the Morning.

Let us adore Jesus Christ, Our Lord, revealing the splendor of His glory upon Thabor, in order to detach us from the world and to make us desire heaven, by showing us how happy we shall be there (Matt. xvii. 4). At this spectacle let us raise our hopes on high and let us conceive great desires in our heart for heaven. There is nothing **more sanctifying.**

FIRST POINT.

The Transfiguration of Our Lord Teaches us to Desire Heaven.

If, in fact, some rays of glory for one moment only have overwhelmed the apostles with so sweet a joy that Peter, being carried out of himself and overpowered with happiness, exclaimed: *"Lord, it is good for us to be **here**; if Thou will, let us make here three tabernacles, one for Thee, one for Moses, and one for Elias"* (Matt. xvii. 4)—what will it be, O Jesus, to see Thee face to face in all the splendor of Thy majesty, in all the brilliancy of Thy glory, and that, not for a few fleeting moments, as upon Thabor, but always, eternally! For eternally we shall contemplate the beauty of Thy face, eternally shall we enjoy Thy ravishing society, not merely in company with Moses and Elias, but together with all the patriarchs, all the prophets, all the apostles, the martyrs, the confessors, and the virgins, not merely in a tabernacle, raised by the hand of man, but in the very city of God. O sweet and glorious hope! O ravishing destiny! It was that which consoled Job in all his sufferings. *"I know,"* he says, *"that my Redeemer liveth and in the last day I shall rise from out of the earth, and I shall be* *clothed again with my skin, and in my flesh I shall see my God. This, my hope, is laid up in my bosom*" (Job xix. 25-27). It was this that made the Apostle so ardently sigh for the dissolution of his body (Philipp. i. 23), and inspired St. Teresa with such longing desires to die: "*O life too long!*" she exclaimed; "*O death too tardy, how my exile is prolonged!*" It was this that made St. Gregory of Nazianzen say: "When I consider the great happiness which we gain in dying and the little we lose in parting with life, I can hardly contain the ardor of my desires; and I say to God: When, O Lord, wilt Thou take me away from earth and bring me to my home?" (Greg. Naz., *Orat. x. in Sacerd.*) Such ought to be the feelings of every Christian. For he, says St. Augustine, will never rejoice in heaven as a citizen who does not sigh here below as an exile. The true Christian, he says elsewhere, suffers at being obliged to live and is delighted to die; life is a cross to him, death an enjoyment. Are these our dispositions? Do we not love exile more than our country, earth more than heaven; and do we not esteem it a happiness to be long exiled from Paradise and to enter into it as late as possible? Oh, what inconsequence is ours! We say to God, Thy kingdom come, and our captivity pleases us, and we seek to establish ourselves in it as if we were to live here always. We are making our way towards happiness and we are in no haste to attain it; we are sinking in the midst of waves and we do not long for the port!

SECOND POINT.

How Useful to the Soul are Holy Desires

1st. They console it in all the trials of life and in all bodily infirmities. What, indeed, are all these trials to a soul kindled with holy desires for Paradise, where it hopes to receive a magnificent compensation? It says to itself, I suffer, it is true, but what is that compared with the happiness which awaits me, and the glory I shall enjoy when my body, transformed into the likeness of the body of my Saviour, shall be clothed with light as with a garment, shining with the splendor of the sun, impassible, immortal? Blessed be the suffering which will be the means of obtaining so much happiness for me! 2d. Holy desires for heaven detach from all temporary things; the soul which is filled with these great hopes sees that the whole world is infinitely below it; it henceforth aspires only to the eternal bliss of Paradise, and it says with St. Ignatius: "*How vile the earth seems when I look at heaven!*" 3d. These holy desires fill the soul with a holy ardor for salvation, and it says to itself, as did St. Augustine, "*If trials make me afraid, may the recompense for them encour-*

age me." When we remember that the smallest trials endured in a Christian manner, the least act of virtue, the least sacrifice, the least prayer well performed, shall have as recompense an eternal weight of glory, there is naught which costs us anything, and we seize joyfully upon all that tends to our salvation. Oh, of how many graces do we deprive ourselves by this forgetfulness of heaven which is so habitual to us! Let us acknowledge it and raise our hearts to heaven. *Sursum corda.*

Resolutions and spiritual nosegay as above.

Thursday in the Second Week.

Summary of the Morrow's Meditation.

We will consider to-morrow: 1st, that in the mystery of the Transfiguration is contained a great lesson upon the love of suffering; 2d, the source of the greatest blessing is found in suffering. We will then make the resolution: 1st, to suffer, without discontent and murmuring, all the contradictions and all the crosses we may meet with; 2d, not to listen to the effeminacy which, by means of too excessive care, endeavors to avoid everything which is disagreeable or inconvenient. Our spiritual nosegay shall be the words of St. Paul to the Hebrews: "*Looking on Jesus the*

Author and Finisher of faith, who, having joy set before Him, endured the cross" (Heb. xii. 2).

Meditation for the Morning.

Let us adore Jesus upon Thabor conversing with Moses and Elias, not respecting the glory with which He was resplendent, but the sufferings which He would endure on Calvary (Luke ix. 31). The mouth speaks out of the abundance of the heart, and as His heart was filled with love for the cross, His mouth delighted to speak of it. Let us thank Him for the great lesson He gives us, and let us beg of Him grace to profit by it.

FIRST POINT.

The Mystery of the Transfiguration Teaches us Love of Suffering.

It seems to us as if Jesus, in the midst of His glory, might have given a truce, during a few moments, to the thought of suffering, but His heart sighed so ardently for the baptism of suffering which was to save the world, that even in the midst of the splendors of Thabor He seemed to be able to speak of nothing else. Jesus, Moses, and Elias conversed together, says Holy Writ, on the terrible sufferings and the cruel death which He was to endure on Calvary. Oh, how well suited is this heavenly conversation to make us understand what we ought to love most on earth. In all circumstances, at all times, in all places, we ought to meditate upon, love, bear the cross, and often speak of it to our heart, as Jesus did upon Thabor to Moses and Elias. St. Peter, whom the Holy Spirit had not as yet enlightened respecting the excellence of the cross, thinks only of present happiness and exclaims: *" It is good for us to be here; let us make three tabernacles, one for Thee, and one for Moses, and one for Elias"* (Luke ix. 33). But the Holy Spirit who recounts this saying immediately corrects the error contained in it, by observing that St. Peter did not know what he was saying (Luke ix. 33). He forgets that to enjoy is the portion of eternity, suffering the portion of this present life (Rom. viii. 17), that everything has its season; that, in order to be seated on a future day upon the throne we must attach ourselves here below to the cross; that, in order to have a share in the glory of the resurrection, we must first of all bear the semblance of death (Philipp. iii. 10, 11); that, lastly, we must pass through many tribulations in order to arrive at the kingdom of heaven (Acts xiv. 21). We should be inexcusable if we were to allow ourselves to fall into a similar error, we who behold this law of suffering written in characters of blood upon the very body of Jesus Christ (I. Pet. ii. 21), we who have seen our divine Saviour surfeited, according to the saying of Tertullian, with the pleasure of suffering for us, and who have heard Him declare by His apostle that something would be wanting to His Passion if He did not suffer in all the members of His mystical body as He Himself suffered in all the members of His natural body; lastly, we, in a word, to whom He has given birth in suffering, who are born by His wounds, and have received upon our heads grace flowing with His blood from His veins so cruelly torn. Children of blood, children of suffering, we cannot save ourselves in the midst of enjoyments. Let us pray to Jesus Christ to make us understand these austere truths and to give us strength to put them into practice.

SECOND POINT.

Suffering is the Source of the greatest Graces to us.

1st. Suffering detaches from this world and obliges the heart to rise to heaven, by means of the discomfort which it makes it experience here below, and which proves to it that it is made for something better than the perishable enjoyments of this world, namely, for eternal bliss. Without suffering, our heart would be lost in the love of present things; suffering alone can break the deceptive charms which incline us towards the earth and make us recognize that God alone is the bed of our repose, that outside Him all is vanity and vexation of spirit. 2d. Suffering purifies virtue, disengages it from all alloy, and makes it enter into that blessed state where God alone is everything to the heart. This is why the more God loves a soul, the less He allows it to remain sleeping for a long time at its ease; He troubles it in its vain enjoyments, and does not permit its heart to be soiled by the current of the waters of Babylon, that is to say, by worldly pleasures. 3d. Suffering strengthens virtue and gives it the character of solidity which alone renders it worthy of God. As long as a soldier has not exposed himself to fire in the battle, his courage is open to suspicion. It is in the same way impossible to count upon an effeminate soul which has not been tried in the crucible of suffering. A contradiction, a loss, a want of the respect due to it, is sufficient to make it murmur and complain. It is a deceptive piety which is only a mockery of true piety, false gold which shines in the sun, but which cannot resist the fire and vanishes in the crucible. The soul which is tried by tribulation, on the contrary, is fashioned to suffering and contradiction, and is accustomed to sacrifice, remains calm amidst the trials of life, kisses the hand of God which strikes it, directs a glance of submission towards heaven, and rejoices even in its trials, in which it sees the guarantee of future happiness. Whatever our fantastic nature of human judgment may cause it to suffer, the inequalities of temper which oppose it, the deceptions of self-love, the disgust to or the fatigue consequent upon labor, it is firm and unshaken, and the more its heart is wounded and made to bleed through contradiction, the happier it is to be able to offer itself to God as a victim marked with the sign of the cross of His well-beloved Son. Are these our dispositions?

Resolutions and spiritual nosegay as above.

Friday in the Second Week.

Summary of the Morrow's Meditation.

In order to conform ourselves to the spirit of the Church, which to-morrow honors the holy winding-sheet, we will consider: 1st, how just is the devotion of this precious relic; 2d, how sanctifying it is. We will then make the resolution: 1st, often to represent to ourselves the holy winding-sheet, bearing the impression of the wounds of Our Saviour and impregnated with His blood; 2d, to excite ourselves by this souvenir to the love of Jesus crucified, to horror of sin, to zeal for salvation, and to the virtues which lead to it. Our spiritual nosegay shall be the words of St. Peter: *"Christ therefore having suffered in the flesh, be you also armed with the same thought"* (I. Pet. iv. 1).

Meditation for the Morning.

Let us adore Jesus Christ descended from the cross and wrapped in the winding-sheet which Joseph of Arimathea had bought; let us venerate His sacred body always united to the person of the Word, and therefore always worthy of the supreme worship of latria. Let us unite ourselves with the adoration which was then rendered to Him by the most Holy Virgin.

FIRST POINT.

How Just is the Devotion to the Holy Winding-sheet.

This devotion dates from the very era itself of Christianity. The gospel, in fact, shows us several winding-sheets carefully folded by the angels in the tomb (John xx. 5). The chief of these winding-sheets, preserved by Nicodemus, passed from his hands to Gamaliel, from Gamaliel to St. James, who transmitted it to St. Simeon, and it was kept in the church of Jerusalem until the year 1137. Carried away at that date by Guy de Lusignan to Cyprus, it was taken from thence to France in 1450 by the widow of the last of the Lusignans, who made a present of it to the Duchess of Savoy. Since that time the royal house of Savoy has kept it until the present day, an object of veneration to the people. God has made known, by many miracles, how acceptable to Him is this devotion, and the Holy See, obeying the indication given by Heaven, authorized to receive the precious treasure a church which Paul II. elevated to the rank of a collegiate church, to which Sixtus IV. gave the title of the Holy Chapel, and where Julius II. permitted the office of the holy winding-sheet to be said. Encouraged by such authorities, the devotion to the holy winding-sheet increased on every side. St. Charles came to oblate his heart in the presence of this venerable relic. Madame de Boissy, before she brought into the world St. Francis de Sales, came there to recommend, with abundant tears, the blessed fruit which she bore in her womb. St. Francis de Sales himself came to Turin to venerate the holy relic, and could not restrain his tears at the sight of the wounds of the Saviour impressed upon the winding-sheet. This devotion of the Church and of the saints has nothing in it which ought to astonish us; for, if we honor the cross as a memorial of the Passion of the Saviour, if a crucifix painted by a skilful hand excites our devotion, how much more ought it to be excited by the representation of the wounds and sufferings of the Saviour, made not by the hand of man, but by the contact of the very body itself of Jesus Christ!

SECOND POINT.

How Sanctifying is the Devotion to the Holy Winding-sheet.

Is it possible indeed for us to represent to ourselves what the holy winding-sheet offers to the eyes of him who contemplates it; that body all bloody, that head crowned with thorns, those feet and those hands pierced with nails, that side opened by the lance—in a word, all the wounds which tore the sacred flesh of the Saviour, from the top of His head down to the soles of His feet, without saying to ourselves: Since my Saviour suffered so much in order to save me, I will not lose the fruit of so many sufferings; since my salvation cost Jesus Christ so dear, I will not lose my soul by refusing to do myself violence infinitely less painful? I will be a saint. At the sight of this holy winding-sheet I detest the sins for which my Saviour shed so much blood, and I embrace the penance which expiates it. Could I be effeminate and sensual after looking at the semblance of this wounded body? Could I shut my heart to the cry which comes forth from the wounds impressed upon this winding-sheet: "*God so loved the world as to give His only begotten Son*" (John iii. 16), and not myself cry from the bottom of my heart: "*Let us, therefore, love God, because God first hath loved us*" (I. John iv. 19). Oh, what a hard heart we must have not to allow ourselves to be touched by so many sufferings endured for love of us (St. Augustine).

Resolutions and spiritual nosegay as above.

Saturday in the Second Week.

Summary of the Morrow's Meditation.

We will to-morrow terminate our meditations upon the Transfiguration by considering: 1st, the profound humility of Jesus Christ which is shown in this mystery; 2d, the universal detachment which this mystery reveals in the apostles. We will then make the resolution: 1st, to attach ourselves to God alone, without desiring anything else; 2d, never to say or do anything through self-love or human respect. Our spiritual nosegay shall be the words of St. Paul: Jesus Christ is everything to the heart (Coloss. iii. 11).

Meditation for the Morning.

Let us prostrate ourselves in spirit at the feet of Jesus transfigured; let us there admire the humility which this mystery reveals to us in Jesus Christ, and the detachment which it reveals in the three apostles who were present upon Thabor. Let us beg of Him to infuse these dispositions into our souls.

FIRST POINT.

The Transfiguration shows forth the Profound Humility of Jesus Christ.

Jesus, in revealing the glory to which His holy humanity has a right in virtue of its hypostatic union with the Word, enables us thereby to understand the profound humility which led Him to keep constantly hidden so magnificent a privilege. It is the only occasion, during the whole course of His life, in which He allows a few rays of His glory to escape; and even then He does it only in order to strengthen the faith and sustain the courage of His apostles in the midst of the persecutions which await them ; it is only in the presence of the three apostles, in a place apart and solitary, in order not to allow what might make Him the object of honor and praise to show itself more than was necessary ; it was only during a few short moments, and immediately afterwards He resumes His poor, humble, and obscure state ; and lastly He recommends His three apostles to keep secret what they had seen, to say nothing to anyone, and to leave Him the whole of His obscurity (Matt. xvii. 9). O admirable humility ! His transfiguration indubitably shows that He has at His disposition riches compared with which gold and precious stones are but as so much dust ; and yet He leads the poorest of lives.

The foxes have holes and the birds of the air a nest, and He has no place where to repose His head! His transfiguration indubitably shows that He is great beyond all thought; that Moses and the prophets are only His servants and His messengers; and yet He hides Himself under the lowest and most humble exterior. He conceals from the eyes of the world all that is glorious in Him; and if, later on, He chooses Jerusalem as the most elevated theatre where He could show Himself, it will be only to suffer there, on the great day, opprobrium and confusion. His transfiguration indubitably shows that He possesses in Himself all the joys of heaven; and yet He delivers up His soul to anguish, His body to suffering, to hunger, to thirst, to fatigue, to suffering, and to death. What lessons of virtue! Let us prostrate ourselves, let us adore, let us love and imitate Him. Let us no longer seek to make a parade of what is an honor to us and to hide what humbles us.

SECOND POINT.

The Transfiguration shows forth a Universal Detachment in the Apostles.

The apostles are so ravished by the beauties which they discover in Jesus that they no longer desire anything else here below. "*Lord,*" they exclaim, "*it is good for us to be here;*" with

Lessons of Humility and Detachment. 177

Thee alone we have all, and the heart has nothing more to desire here on earth. We have in the world relations, friends, acquaintances, a thousand things to which we cling; but, Lord, in Thee we have everything; for Thy sake we heartily consent to abandon everything; we esteem ourselves to be rich enough if we possess Thee; sufficiently happy if Thou art with us, honored enough if we are in Thy company. Let us remain here (St. Ambrose). It is thus that a soul which has appreciated Jesus, which has studied His beauties and His charms, is detached from all created things, says St. Ambrose. Prosperity does not intoxicate it, adversity does not cast it down; whether it be praised or blamed, whether it be honored or despised, whether it be rich or poor, it signifies little. Jesus alone is its all. Like the apostles upon Thabor, it sees nothing but Jesus in all things; it seeks only to please Jesus, it aspires to nothing but the esteem and love of Jesus; and with the eyes of its heart fixed upon Jesus, all the rest is nothing to it. Wherefore, it says, should I attach myself during life to that which death will take away from me? wherefore love during time what will be nothing to me during eternity? Is it thus that our heart is detached from all that passes away, and is fixed upon Jesus, who does not pass away?

Resolutions and spiritual nosegay as above.

Third Sunday in Lent.

The Gospel according to St. Luke, xi. 14–28.

"At that time Jesus was casting out a devil, and the same was dumb; and when He had cast out the devil, the dumb spoke, and the multitude were in admiration at it; but some of them said: He casteth out devils by Beelzebub the prince of devils. And others tempting asked of Him a sign from heaven. But He, seeing their thoughts, said to them: Every kingdom divided against itself shall be brought to desolation, and house upon house shall fall; and if Satan also be divided against himself, how shall his kingdom stand? Because you say, that through Beelzebub I cast out devils. Now if I cast out devils by Beelzebub, by whom do your children cast them out? Therefore, they shall be your judges. But if I by the finger of God cast out devils, doubtless the finger of God is come upon you. When a strong man armed keepeth his court, those things are in peace which he possesseth; but if a stronger than he come upon him and overcome him, he will take away all his armor wherein he trusted, and will distribute his spoils. He that is not with Me, is against Me; and he that gathereth not with Me, scattereth. When the unclean spirit is gone out of a man, he walketh through places without water, seeking rest; and not finding, he saith: I will return into my house whence I came out; and when he is come, he findeth it swept and garnished. Then he goeth and taketh with him seven other spirits, more wicked than himself, and entering in they dwell there; and the last state of that man becomes worse than the first. And it came to pass, as He spoke these things, a certain woman from the crowd, lifting up her voice, said to Him: Blessed is the womb that bore Thee, and the paps that gave Thee suck. But He said: Yea rather, blessed are they who hear the word of God, and keep it."

Summary of the Morrow's Meditation.

As the gospel of to-morrow speaks to us of the sin of backsliding, we will consecrate our next meditation to the consideration of this sin being: 1st, injurious to God; 2d, terrible for man. We will then make the resolution: 1st, every day, after our examination of conscience, to fix upon the fault which it is the most necessary for us to correct, in order to avoid above all in regard to these faults the sin of backsliding; 2d, to urge ourselves to lead a better life during the day before us than we did during the one which preceded it. Our spiritual nosegay shall be the words of the gospel: "*The last state of that man is made worse than the first*" (Matt. xii. 45).

Third Sunday in Lent.

Meditation for the Morning.

Let us adore Our Lord teaching us in the gospel of to-day the gravity of the sin of backsliding. The state, He says, of him who, after having been delivered from the devil, returns and places himself again under his rule, becomes worse than the first. Let us thank Him for so important a warning and ask of Him grace thoroughly to profit by it.

FIRST POINT.

How Injurious to God is the Sin of Backsliding.

1st. It is horrible ingratitude; God had pardoned us our preceding falls, and this pardon gave Him a right to our greatest gratitude. Ought we not indeed to be abundantly thankful for His generosity which forgot our sins, for His grace which effaced them, for the blood of Jesus Christ which merited this grace for us, for the recovery of our rights to Paradise and the eternal possession of God, lastly for the gratuitousness of so great a favor, since not only had we not merited it, but we had become supremely unworthy of it through sin! And yet, instead of thanking Him and of blessing Him for so much love, we recommence our offences; we care so little for the loss of His grace, or, if our fault is only venial, for the diminution of that same grace! What shameful and guilty ingratitude! O Christian soul, how this backsliding degrades and lowers thee! (Jer. ii. 36.) 2d. It is an unworthy abuse of the goodness and patience of God. Because God is good, we do not trouble ourselves about not committing sin. He has pardoned me this sin, we seem to say, He will pardon me again if I commit it a second and a third time. And we fall again. O man! exclaims St. Paul, how canst thou thus despise the riches of the goodness of God, of His patience, and of His longanimity? (Rom. ii. 4.) Dost thou not understand that the great goodness of God is a reason for serving Him better, and that to make of it a motive for offending Him is to amass treasures of anger upon thy head? (Rom. ii. 4, 5.) 3d. It is shameful perfidy. Each time that we approach the Holy Tribunal or the Holy Table, we protest that we will not fall again; the blood of Jesus, which is applied to us by absolution or given to us in communion seals our protestations; and yet at the least opportunity which presents itself of pleasing ourselves or others, we violate these sacred promises! Is it not shameful perfidy? O my God, pardon, mercy! (Ps. l. 3.)

SECOND POINT.

How Terrible the Sin of Backsliding is for Man.

1st. This kind of sin weakens us. By familiarizing us with evil, it diminishes the horror of it, consequently, it diminishes the will to resist it; by making us fall, it injures our strength in proportion to the height whence we have fallen and the depth to which it has abased us. How, before our fall, we were raised so high! We were the friends of God; and through our backsliding we have descended so low, even as low as hell if the sin be mortal, and as low as purgatory if it be venial. Oh, how such backsliding bruises and weakens us! 2d. Each backsliding increases the difficulty of rising again. *"It is impossible,"* says St. Paul, *"for those who were illuminated . . . and are fallen away to be renewed again to penance"* (Heb. vi. 4, 6). Doubtless this impossibility is not to be understood literally; as long as man breathes salvation is possible to him; but at least it indicates a serious difficulty which ought to make us tremble. Backslidings, in fact, attach the will to evil, and engender the habit which soon becomes a second nature, in such a manner that we will no longer either take means for rising again, or put away the obstacles which hinder our return, or combat our bad inclinations. The small effect which certain efforts have produced disgusts us from making fresh ones, and makes us believe that it is impossible for us to correct ourselves. Lastly, the shame of having made so many attempts without success keeps us back, and we remain always in the same state

Such is the kind of impossibility with which St. Paul threatens the sinner who backslides. Is there not matter here for making us tremble and inspiring us with a firm will not any more to commit the sin of backsliding?

Resolutions and spiritual nosegay as above.

Monday in the Third Week.

Summary of the Morrow's Meditation.

We will resume, to-morrow, our meditations on the Sacrament of Penance, interrupted by the gospels so full of interest on which we have been meditating ; and we shall see that we must bring to our confessions : 1st, a really interior contrition ; 2d, a really universal contrition. We will then make the resolution : 1st, to make, every evening, after our examination of conscience, an act of interior and universal contrition ; 2d, every day or during the night when a fault escapes us to make an act of interior contrition. Our spiritual nosegay shall be the words of the Psalmist : *" A sacrifice to God is an afflicted spirit : a contrite and humbled heart, O God, Thou wilt not despise"* (Ps. l. 19).

Meditation for the Morning.

Let us adore Our Lord in the Garden of Olives, seeing from thence, clearly and distinctly, the sins of all ages, of which He had assumed the expiation. The sight throws Him into mortal anguish; He weeps over the offence committed against God and the ruin of man, not only with tears in His eyes, but with the blood of His body; He weeps in all His members, St. Bernard says, and inundates the ground with tears of blood (Luke xxii. 44). Let us have compassion on the afflicted Saviour; let us weep with Him, for it is over our sins that He weeps (*Serm.* iii. *in Nativ. Dom.* 4).

FIRST POINT.

We must Bring a Really Interior Contrition to our Confessions.

Jesus Christ, that perfect model of contrition, in the Garden of Olives clearly teaches it to us; His heart feels so acutely sorrow for sin, that He is sorrowful even unto death. Besides, reason itself teaches us the necessity of this interior contrition. Since it is the heart which has offended God, it is the heart which ought to make reparation for the offence and to be bruised with grief at having displeased a God so good and so worthy of being loved. God only pardons in so far as the heart repents to such an extent as to make it wish that not for the whole world had it committed the faults which it deplores. "*Rend your hearts*" (Joel ii. 13), says God to sinners, "*make to yourselves a new heart*" (Ezech. xviii. 31). God beholds, not the eyes which shed tears, nor the lips which pronounce formulas, but the heart which has a sincere horror for sin committed (I. Kings xvi. 7). In vain, then, may the mouth articulate acts of contrition ; in vain may the mind and the imagination form within us an idea of sin to the extent of making us persuaded that we are contrite ; in vain shall we utter sighs and groans, and shed tears, and make long prayers, and protestations of renunciation of sin ; all will be of no use if, at the bottom of our hearts, we do not sincerely regret having offended God, if we have not a real detestation, a pronounced hatred of sin, with a sincere sorrow for having committed it. Let us here examine ourselves in presence of the Lord ; do we bring to our sins a really broken heart at having offended God, saying to Him with St. Bernard : *"How can I dare raise my eyes to Thee, I who am so wicked a son of so good a Father"* (*Serm.* xvi. *in Cant.*). Instead of sincerely deploring our faults, have we not refused to acknowledge them and sought to disguise them in our own eyes and the eyes of our confessor, by covering them over with excuses, in order not to have to blush for them, justifying our evil tempers and our impatiences by the wrongs others have committed against us, our backbitings and our criticisms by the unreasonable conduct of our neighbor?

SECOND POINT.

We must Bring to our Confessions a Really Universal Contrition.

That is evident when mortal sins are in question; if there be a single one which we do not detest sincerely and from the bottom of our soul, our contrition is worth nothing, our confession is sacrilege. God does not love a heart which loves sin, which essentially displeases Him; and it is making a mockery of God to say to Him, *I love Thee,* when we have an affection for that which He supremely detests. If venial sins be in question, the contrition is not null because it may not be universal, because as venial sin only weakens God's friendship for us without destroying it, we can repent of some without repenting of others; nevertheless there results from it several serious injuries to the soul; 1st, sins for which we preserve affection are not remitted, and remain in the soul like a hideous spot, which disfigures it, which cools the friendship of God for it and diminishes His graces; 2d, the absolution not being applied to these sins, does not confer the grace of correcting them, and does not produce in the soul that fulness of justification which a heart entirely devoted to God would have obtained. Let us here examine ourselves and see whether there are not within us certain favorite sins which we do not sincerely wish to renounce, certain attachments which we will not break off, certain faults to which we have a greater tendency, which give us more pleasure, and for which we have not frank contrition.

Resolutions and spiritual nosegay as above.

Tuesday in the Third Week.

Summary of the Morrow's Meditation.

We will consider in our next meditation two other essential characters of contrition; and we shall see that it ought to be: 1st, supreme; 2d, supernatural. We will then make the resolution: 1st, to re-awaken in our soul faith in these two truths, and to maintain the habitual sentiment of them in our soul; 2d, to make more resolute acts, every evening at our examination, and each time that we confess. Our spiritual nosegay shall be the words of the Psalmist: *"I have hated and abhorred iniquity"* (Ps. cxviii. 163).

Meditation for the Morning.

Let us adore Jesus overwhelmed with grief in the Garden of Olives (Matt. xxvi. 37; Mark xiv. 33); He sees the frightful evils produced by sin; hell opened, paradise closed, God despised, the devil upon the throne; and the sight of all this **saddens** His soul to such an extent that it is necessary for an angel to come down from heaven and sustain Him (Luke xxii. 43). Let us render to His soul filled with desolation all the homage of which our hearts are capable.

FIRST POINT.

We must Bring to our Confessions a Supreme Contrition.

A supreme contrition is that which makes us feel more sorrow for having offended God than for all the evils which could possibly happen to us. And what is more just, O my God, than such grief? Dost Thou not deserve to be loved above all things? Can the loss of fortune and reputation, even the deaths of relatives and friends, be weighed against the loss of Thy grace and Thy friendship, with the loss of heaven throughout eternity, which is the consequence of sin? No, doubtless; the least particle of good sense assures us of it. It is not necessary that the sorrow of having sinned should be as sensible as the grief of having lost a father or a mother; God does not ask sensibility from us, because it does not depend upon ourselves; but He requires that we should detest sin as a supreme evil, and that we should be ready to lose all and suffer all rather than commit it one single time. Nor is it opportune that we should represent to ourselves all kinds of ills, such as the tortures of martyrs, in order to ask ourselves if we are ready to bear them rather than commit sin; for we do not actually possess the grace necessary for such a trial; it suffices to say: "If I were in such a case, I would pray to God with all my heart to give me the necessary grace; I am confident that He would not refuse it, and this confidence gives me courage to say: Rather all evils than sin." Let us examine ourselves if we have brought to our confessions this supreme contrition.

SECOND POINT.

We ought to Bring to our Confessions a Supernatural Contrition.

If, in fact, our contrition were purely natural in its principle, it could have no value in the supernatural order. Our nature cannot of itself rise to the supernatural order; we can do nothing of ourselves, says St. Paul; we can neither have a thought useful for salvation nor say a single meritorious word. It is then from Thee, O Divine Spirit, that we ask true contrition, and it is from Thee alone that we can obtain it, but on one sole condition: it is that we should base it upon supernatural motives as its principle. If we detest sin only because it has rendered us unhappy, tormented us with remorse and disquietude, ruined us in our fortune, our health, or our reputation, it would be a vain and sterile contrition. Useful contrition has higher views; through it the soul, borrowing its motives from faith, holds sin in supreme horror, and feels a profound regret for having committed it, because in committing it it renounces the friendship of God and its portion in paradise, it gives itself to the devil and exposes itself to eternal damnation, it incurs the hatred and the malediction of its heavenly Father, it has been the cause of the Passion of Jesus Christ, of His mortal anguish in the Garden of Olives and of His agony on the cross; but above all because it has displeased God whom it loves above all things, because it has offended His infinite majesty, outraged His goodness and His love. This is what renders the soul inconsolable for its faults, this is what breaks its heart and humbles it beyond all power of speech (Ps. l. 19). O Jesus, crucified for my sins, Thou alone canst infuse these sentiments into me; let fall upon my heart some drops of Thy blood to soften it; speak to it by all Thy wounds as by so many mouths; and may these wounds produce in me the supernatural contrition which purifies the soul and inclines it to live henceforth only for Thee, to love only Thee! Let us here examine ourselves and see whether we have brought to our confessions a really supernatural contrition in its principles and in its motives.

Resolutions and spiritual nosegay as above.

Wednesday in the Third Week.

Summary of the Morrow's Meditation.

As contrition, in order to be valid, ought to be based upon motives of faith, as we saw in our last meditation, we will meditate to-morrow on the first of these motives, and we shall see: 1st, how sin, being an offence against God, is an evil which merits tears; 2d, how greatly the circumstances in which the sinner commits it renders it still more horrible. We will then make the resolution: 1st, to be thoroughly penetrated before we present ourselves at the holy tribunal with this great motive of contrition; 2d, to recall it to ourselves every day, in the morning and in the evening, in order to excite in us horror for sin. Our spiritual nosegay shall be the words of the prodigal son: *"Father, I have sinned against heaven and before Thee; I am not worthy to be called Thy son"* (Luke xv. 18, 19).

Meditation for the Morning.

Let us adore Our Lord Jesus Christ prostrate on His knees before the majesty of His Father (Luke xxii. 41). In this humble posture He asks of Him pardon for the outrages committed against Him by sin; He offers reparation for them, and consents to bear their penalty. Let us unite ourselves with the sentiments of His afflicted heart, and ask of Him a share in these holy dispositions.

FIRST POINT.

How Sin, inasmuch as it is an Offence against God, is Worthy of all our Tears.

Alas! my God, if I had only failed in matters of pure counsels, it would have been enough to merit all my tears; for is it not a very deplorable irreverence when Thou sayest: "Do this, it will be more agreeable to Me; do not do that, thou wilt displease Me," to have the impudence not to yield to the authority of Thy desires, and only to obey when Thou hast the rod in Thy hand, like a vile slave who obeys only when he hears a voice menacing him, or like a false friend, who does not respect the desires of his friend, and is not afraid of displeasing him. The impudence is much more shameful still, O my God, when, passing from counsels to orders, Thou sayest: "Do this, I command thee; do not do that, I forbid thee; if thou dost not obey, the fires of purgatory will execute justice on thy rebellious will." And yet I have the audacity to do what Thou forbiddest, and to omit what Thou commandest. What, Lord! I, a vile creature, a worm of the earth, whom Thou mightest annihilate at a glance, and whom Thou dost preserve through pure mercy—

I disobey Thee—I who desire that all should bend to my will, and am angry if my servants do not promptly execute my slightest orders! I disobey Thee to Thy face; whilst seeing, through faith, the majesty of Thy eyes fixed upon me, I do, before Thy eyes, what often I would not do before the eyes of a servant, and that not once, but thousands of times, and daily! Is not that a fault which calls for all the tears I can shed? And yet that is only venial sin. What, then, O my God, is mortal sin? Ah, if I had committed but one in my whole life it would have been enough to make me pass all the rest of my life in shedding contrite tears. At least, in venial sin I did not entirely renounce Thy friendship; I did not exchange my right to heaven for hell; but I see that mortal sin makes me break entirely with Thee, incur Thy hatred, make myself a butt for Thy great anger, and yet I hold it all of no account! If I thought that in sinning I should displease the world as much as Thee; that I should inflict an injury upon my honor, my fortune, my pleasures, as well as my innocence, I should take good care not to commit it; but because, in sinning, I offend only Thee, and that it is only Thy friendship I lose, I allow myself to sin! Pardon me, O Lord, for showing Thee such contempt. I see Thee displaying before me, if I sin, all the power of Thy vengeance, all the eternity of Thy chastisements, and yet I sin, spite of Thy threats! I see that Thou askest of me only what is infinitely just, which my conscience dictates to me and my reason approves, and I despise Thy orders, in spite of my reason and in spite of my conscience! I put in the balance against Thee a passing gratification, a soiled and tainted pleasure, which enters into the soul only in order to take with it unhappiness and remorse; and yet, in this alternative, passion gets the upper hand; dirt is preferred to Thee! O crime! O subversion! O abyss of iniquity! Pardon, Lord, mercy!

SECOND POINT.

How the Circumstances in which the Sinner Offends God render his Fault still more Horrible.

1st. There is the treachery of it. Because, at my baptism and in my many confessions and communions, I made an oath of fidelity to Thee, O my God, and behold, after having taken upon me so many engagements, I have, nevertheless, been unfaithful to Thee! O faith of treaties, where art thou? O violated oaths! O chief of felonies! O disloyal Christian! O traitor and perjurer! 2d. There is the ingratitude of it. Jesus Christ died for me; He has given Himself to me in the sacraments; He has pursued me with His graces, and His love has surrounded me day and night with His natural and supernatural graces; and I, who am overwhelmed with His favors, I have turned against Him; I have employed His own gifts, my intelligence, my will, my senses, in offending against Him! O horrible ingratitude! 3d. There is in it the rebellion of the subject against his sovereign; of the son against the best of fathers; of the friend against the most faithful of friends; of the creature against the Creator; of weakness against omnipotence; of littleness against infinite greatness! There is in it even more than all this; there is the crime of high treason against the divine Majesty; there is in a certain way a double deicide: the first, in that my sins, the cause of the death of Jesus Christ, are as the executioners who nailed Him to the cross by a crime worse than that of the Jews, who would not have crucified the Saviour had they known Him; and yet I, who knew Him, have crucified Him! The second deicide consists in the sinner desiring that God should not know his sin; and supposing that He knows it, he desires that God should not detest it; and supposing that He detests it, he desires that God should not punish it. Now, to desire all this is to desire that God should be deprived either of His knowledge or of His holiness or of His power. It is, consequently, to desire that God should not be God. What a horror! Oh, how hateful, then, is sin! and what a firm resolution we ought to make to avoid it a thousand times more than the greatest evils which could happen to us!

Resolutions and spiritual nosegay as above.

Thursday in the Third Week.

Summary of the Morrow's Meditation.

We will meditate to-morrow on a second motive for contrition; it is the supreme displeasure which, 1st, venial sin; 2d, mortal sin cause to God. We will then make the resolution: 1st, very carefully to avoid the least venial sins, since God has such a great horror of them; 2d, to weep, all the days of our life, over the mortal sins which we have had the unhappiness to commit during the past. We will retain as our spiritual nosegay the words of the Psalmist: "*My sin is always before me*" (Ps. l. 5).

Meditation for the Morning.

Let us prostrate ourselves with trembling before the justice of God pursuing sin with implacable hatred (Wis. xiv. 9). O God, Thy justice is higher than the mountains and deeper than the abysses (Ps. xxxv. 7); it surpasses all imagination. I adore it without understanding it, but at the same time I love it, because everything in Thee is amiable! Be Thou forever praised and blessed in Thy justice as well as in Thy goodness.

FIRST POINT.

How we ought to Weep over Venial Sins, inasmuch as they Displease God.

God hates venial sin so much that in the next life He visits it with chastisements which, during almost an eternity, are a kind of hell, and He keeps the gates of His Paradise closed against souls which are His friends and are dear to Him until the complete expiation of the least of their sins. He hates it so much that even in this life He has often visited it with terrible chastisements. The wife of Lot permitted herself to indulge in thoughtless curiosity; at that very instant she is struck dead (Gen. xix. 26). A man is discovered picking up a little wood on the Sabbath day; stone him and let him die, said the Lord (Num. xv. 32 *et seq.*). Moses indulged in a little mistrust of God; he is not allowed to enter into the promised land, which he had deserved to see by forty years of service (Deut. i. 37). A prophet, through complaisance, remains a little longer than necessary in the place to which he had been sent; a lion comes out of the forest and devours him (III. Kings xiii. 22, 24). David, animated by secret vanity, causes his people to be numbered;

seventy thousand men die of the pestilence (II. Kings xxiv. 15). O God, what then is venial sin in presence of Thy divine Majesty? How bitterly ought we to weep over evil which displeases Thee so supremely! and how just it is to bring every time to the holy tribunal a lively contrition for our sins, accompanied with a firm resolution to correct ourselves of them. Is it thus that we act?

SECOND POINT.

How much we ought to Weep over Mortal Sin, inasmuch as it is Supremely Displeasing to God.

When we reflect on the terrors of hell, and when we call to mind that those who are suffering such incredible torments there were the children of God, His well-beloved, for whom He had given the whole of His blood; and that a single mortal sin, converting such ineffable love into such implacable anger, will make the whole weight of His divine vengeance weigh them down throughout eternity, we are seized with stupor, and we exclaim: Oh, how much, then, does mortal sin displease Thee, O my God, and with what hatred dost Thou pursue it! If from hell we raise our thoughts to heaven, what do we see? Empty places which were formerly occupied by angels, pure spirits, shining with admirable beauty, clothed with the most magnificent perfections, masterpieces of the hand of God. A day comes when they allow themselves to indulge in a proud thought; at that very moment God pronounces against them a terrible sentence. But, O Lord, if Thou wouldst grant them pardon, they would praise Thee throughout eternity; if Thou wilt cast them into hell they will blaspheme Thee everlastingly, and will drag down to eternal damnation millions of men; it does not signify: let them fall into the bottom of the abyss. But they have only committed one single sin; it is their first sin, and, after all, it is only a sinful thought; it does not signify: let them fall into the bottom of the abyss. O holiness of my God, how pitiless is thy hatred of sin! But if Thou dost thus punish the officers of Thy court, what ought not I to fear—I, the last of Thy servants, guilty of a thousand treasons; I who have sinned, not only once and in thought, but millions of times and in all my senses, all the members of my body, in all the powers of my soul, and against the majority of Thy commandments (St. Bernard). From heaven, thus depopulated of a portion of its inhabitants, I descend to the terrestrial paradise, and I there see the place which Adam occupied when he was innocent. A day came when he had the misfortune of yielding to an intemperance, which seems apparently very slight; he ate a certain fruit which God had forbidden him, and immediately he lost all the graces of his first state; he was condemned to all kinds of evils, even to death, and not only he, but all his posterity. All men, down to the end of the world, will be a prey to innumerable miseries, to war, to pestilence, to famine, to murders, to tempests, to ignorance, to concupiscence; nay, every one of them would have been damned if the entirely gratuitous mercy of God had not redeemed us. Great God! how many punishments at one and the same time for a single sin! and if one single sin displeased Thee to such a degree as to make Thee resolve to visit the world with so many calamities, what will be the consequence of my innumerable sins? Can I ever weep enough for them, and conceive a sufficiently lively contrition for them? Nevertheless, my God, it is not even in these things that is shown in all its intensity the horror which Thou hast for sin. I take my crucifix in my hand, and I say to myself: He whose image I contemplate was the only and well-beloved Son of God; He was God, but because He took upon Himself the semblance of sin, His heavenly Father launched upon His head all the weight of His anger; He delivered Him up to the most cruel of torments, the most terrible ignominy, to death, and death upon the cross. O sin! how horrible thou art in the sight of God; how I ought to regret and to weep over the evil I committed in allowing thee to enter into my heart! If, for the mere semblance of sin, God thus treated His only Son, how for so many real sins will He treat a rebel and contemptible subject like me? If wood that was green had to pass through such a furnace, what will it be with dry wood which is ready to be consumed by fire? (Luke xxiii. 31.) Behold here the most powerful of motives for weeping over sin and conceiving a better contrition for it.

Resolutions and spiritual nosegay as above.

Friday in the Third Week.

Summary of the Morrow's Meditation.

We will meditate to-morrow upon the devotion to the five wounds of Our Lord, which the Church presents to us as the object of our devotion, and we will consider: 1st, that nothing is more just than this devotion; 2d, that the most precious graces are attached to it. We will then make the resolution: 1st, to keep a crucifix before our eyes during our work, and to look at it lovingly, especially in our trials and temptations, and often to press our lips upon its venerable wounds, above all upon the wound in the sacred side; 2d, to practise some mortification in honor of the five wounds. Our spiritual nosegay shall be the words of Isaias: "*He was wounded for our iniquities, He was bruised for our sins*" (Is. lii. 5).

Meditation for the Morning.

Let us prostrate ourselves before the cross of our Redeemer, and let us render our homage of adoration, thanksgiving, and love to the two wounds of His feet, to the two wounds of His hands, but above all to the wound of His sacred side. Oh, how worthy of veneration are these wounds, and how just that our hearts should be filled with love in contemplating them! O sacred wounds! I cannot honor you as much as I would, but I offer you the sentiments of Mary and St. John at the foot of the cross. I have a right to do this, since Mary being my mother and St. John my brother, their merits are a heritage of which I can dispose in my favor.

FIRST POINT.

Nothing is more Just than the Devotion to the Five Wounds.

A son would not be looked upon as a man but as a heartless monster who could behold with indifference and without any emotion of compassion, of gratitude and of love, the wounds which his father had received in order that he might save him from the greatest possible misfortune, and at the same time to procure him the greatest blessings. Such, and worse still, would be the Christian who would feel nothing but indifference for the wounds of the Saviour, because Jesus Christ received them to save us from hell and to open heaven to us; to offer us in them so many sources of salvation, whence we may derive grace, strength, and consolation (Is. xii. 3). O Christian soul, exclaims St. Bonaventura, how, at the remembrance of these wounds, canst thou contain thy transports? Our amiable Jesus makes a great wound in His feet and His hands in order to receive thee and thou dost not hasten to enter it. He has opened His side in order to give thee His heart, and thou dost not hasten to unite thyself heart to heart with Him. Ah! as for me, says the holy doctor, it is there that I delight to dwell (Matt. xvii. 4), it is there that I will make three tabernacles, the first in the hands of my Jesus, the second in His feet, the third in His sacred side. It is there that I will take my rest; it is there that I will watch, that I will read, that I will converse. O most amiable wounds, the eyes of my heart will always be fixed on you; during the day from the rising of the sun to its going down, and at night as many times as sleep shall withdraw from my eyelids. I will above all keep myself within the opening of the sacred side in order there to speak to the heart of my Master and to obtain from it all that I desire. O Jesus, St. Bernard says in the same sense, Thy side is pierced to give us an entrance into Thy heart and to reveal to us by means of this visible wound the invisible wound of Thy love. I will apply my lips to it and I will suck from it the honey of love and the unction of divine consolation (*Serm.* iii. *de Pass. Domini*). Should we be the children of the saints if, after such examples, we had not a tender devotion towards the five wounds?

SECOND POINT.
Graces attached to the Devotion to the Five Wounds.

The soul finds in these wounds all that is necessary to salvation (St. Bernard, *Collat.* 7). I have found nowhere else, says St. Augustine, so efficacious a remedy for all the ills of the soul (*Manual.*, c. xii.). Whatever may be our spiritual maladies, adds St. Bernard, an assiduous meditation upon the wounds of the Saviour will cure them (*Serm.* lxii., *in Cant.* iv. 7). Jesus Christ Himself says by His prophet: They shall look upon My wounds, and they shall be converted (Zach. xii. 10). The heart of Jesus is an ocean, and His wounds are the channels through which flow the waters of grace and mercy (*In Cant.* lxi.), St. Bernard also remarks. It is, in fact, in these wounds that a lively faith is formed (John xx. 27); it is there that confidence in God dilates (*In Cant.* lxi.); it is there, above all, that charity is kindled as at its true source. By dint of considering the excess of love which opened these wounds for us, vile creatures and miserable sinners as we are, the heart is set aflame, and we can no longer live except by love. Therefore St. Augustine called these sacred wounds his refuge in troubles, his asylum in tribulations, his remedies for the infirmities of the soul; it was therein that St. Thomas Aquinas derived all his knowledge; therein that St. Francis of Assisi, by dint of meditating upon them, became, through the seraphic ardor of his charity, a miracle of resemblance to the crucified Jesus; there that St. Bonaventura filled himself with the spirit of piety which embalms all his writings—worthy disciple of St. Francis that he was, and who wore out the feet of his crucifix by kissing them so often, and who never ceased to exhort the faithful to enjoy the ineffable delights and the delicious unction of piety attached to the devotion to the sacred wounds (St. Bonaventura, *Stim. Am.*, c. i. p. i.) If you cannot, says the Imitation of Christ, raise yourself to lofty contemplations, remain humbly in the wounds of the Saviour; you will there find consolation and strength (II. Imit. i. 4). Are these really our dispositions?

Resolutions and spiritual nosegay as above.

Saturday in the Third Week.

Summary of the Morrow's Meditation.

We will employ our meditation to-morrow: 1st, in returning to the first half of Lent through which we have already passed; 2d, in reflecting upon the means for spending in a better manner the second half of this holy season. We will then make the resolution: 1st, to apply ourselves to the practice of recollection and to a spirit of prayer by the frequent use of pious ejaculations; 2d, better to put in practice the instructions given to us and the spiritual reading to which we shall devote ourselves. Our spiritual nosegay shall be the words of St. Augustine: *"Be afraid of losing the grace which is passing by."*

Meditation for the Morning.

Let us adore Jesus Christ, alone in the desert during the holy forty days of which we are celebrating the memory. The divine Solitary calls upon us to become better during this season of salvation. Let us be confounded at having until now responded so ill to His appeal, and let us ask of Him grace to respond to it better during the second half of Lent.

FIRST POINT.

We were not what we ought to have been during the First Half of Lent.

In order to understand it, it is sufficient to consider what we ought to be and what we have been. First, what we ought to be. It is a great error to suppose that in order to assure our salvation it suffices not to commit great crimes. The young man in the Gospel who had kept all the commandments (Matt. xix. 20) refused to embrace the highest perfection, which was to sell all his goods in order to give the price to the poor, and that was enough to make Our Lord exclaim with a sigh: "*How difficult it is for the rich to be saved,*" and for the apostles to ask: "*If he is not saved, who then can be saved?*" (Ibid. 25)—two sentences which seem to prophesy the loss of the unhappy young man's soul. The apostles even had a discussion amongst themselves arising from self-love, but which did not exceed the limits of venial sin (Luke xxii. 24), and yet Jesus Christ said to them: If you are not converted, neither shall you enter into the kingdom of heaven (Matt. xviii. 3). The Bishop of Ephesus, who it is believed was St. Timothy, deserved to be praised by Our Lord for his labors and his zeal; nevertheless he would not have been saved if he had not endeavored to become better. You were more fervent at the beginning, Jesus Christ said to him; "*Be mindful, therefore, from whence thou art fallen, and do penance, and do the first works, or else I come to thee, and I will move thy candlestick out of its place*" (Apoc. ii. 5), that is to say: I will withdraw the light of My grace. All these examples show us clearly that we are mistaken in thinking our salvation to be assured if we do not commit great crimes. In order to make our vocation and our election certain we must take it to heart to lead a perfect life and to multiply our good works (II. Pet. i. 10). We must correspond with the graces which we receive and lead a life in harmony with them, for more shall be demanded from him who has received much (Luke xii. 48). These, then, are what ought to have been our efforts during every day of the first half of Lent. Now is it thus that we have lived? Have we really taken to heart the great work of our perfection? Have we understood that those words of Our Lord, "*Be ye perfect, as your Father who is in heaven is perfect,*" are not the enunciation of a mere counsel, but a precept to tend to perfection according to our strength and the grace given to us by God? Have we consequently endeavored every day to do better than we did the preceding day, and at every hour to do better than during the hour which preceded it? What fruit have we derived from all the means of salvation during this holy season, so many instructions and exhortations, so many spiritual readings and so many pious examples, so many good thoughts and pious emotions, lastly, so many interior and exterior graces? Alas, let us confess, with sighs, that we have not been that which we ought to be.

SECOND POINT.

Means for Spending the Second Half of Lent in a Better Manner.

1st. We must renounce a life which is given up to levity, in order to devote ourselves to the practice of recollection, without which any kind of virtue is impossible. 2d. We must say, from the bottom of our hearts, I am determined to be a saint; and in consequence of this resolution, we must carefully avoid even venial faults, without ever permitting ourselves deliberately to commit any; then we must often put to ourselves this question: Is it thus that the saints thought, acted, prayed, conversed? and regulate our conduct thereby. 3d. We must not resist any grace, but put ourselves in the hands of God, in order to allow ourselves to be led by His Holy Spirit, like a child by the hand of its mother. When we read, or when we have an instruction given to us, we must say, what fruit shall I derive from it? To each good thought which comes to us we must answer God, as Samuel did. "*Here I am, Lord,*" and follow the inspiration. 4th. We must fix upon some special defects, the reformation of which we must pursue during the whole remainder of Lent, such as self-love, our temper, or sins of the tongue.

Resolutions and spiritual nosegay as above.

Fourth Sunday in Lent.

The Gospel according to St. John, vi. 1–15.

"At that time, Jesus went over the sea of Galilee, which is that of Tiberias; and a great multitude followed Him, because they saw the miracles which He did on them that were diseased. Jesus therefore went up into a mountain, and there He sat with His disciples. Now the pasch, the festival-day of the Jews, was at hand. When Jesus therefore had lifted up His eyes, and seen that a very great multitude cometh to Him, He said to Philip: Whence shall we buy bread, that these may eat? And this He said to try him, for He Himself knew what He would do. Philip answered: Two hundred penny-worth of bread is not sufficient for them, that every one may take a little. One of His disciples, Andrew, the brother of Simon Peter, saith to Him: There is a boy here that hath five barley loaves and two fishes; but what are these among so many? Then Jesus said: Make the men sit down. Now there was much grass in the place. The men therefore sat down, in number about five thousand. And Jesus took the loaves; and when He had given thanks, He distributed to them that were sat down. In like manner also of the fishes, as much as they would. And when they were filled He said to His disciples: Gather up the fragments that remain, lest they be lost. They gathered up therefore, and filled twelve baskets with the fragments of the five barley loaves which remained over and above to them that had eaten. Now those men, when they had seen what a miracle Jesus had done, said: This is of a truth the prophet that is to come into the world. Jesus therefore, when He knew that they would come to take Him by force and make Him king, fled again into the mountain, Himself alone."

Summary of the Morrow's Meditation.

We will meditate to-morrow: 1st, on the goodness of Jesus Christ in the multiplication of the material bread which nourishes the body; 2d, on His still greater goodness in the multiplication of the eucharistic bread which nourishes the soul. We will then make the resolution: 1st, to unite with all our repasts a great feeling of gratitude towards Providence who gives them to us; 2d, to honor the Holy Eucharist by frequent and more fervent communions and by more regular and more recollected visits to the Blessed Sacrament. Our spiritual nosegay shall be the words of the

Psalmist: "*How good is God... to them that are of a right heart*" (Ps. lxxii. 1).

Meditation for the Morning.

Let us adore the tenderness of Jesus Christ towards the people who followed Him into the desert; His loving heart is moved by their needs, and He provides for them in a wholly miraculous manner. Let us adore His goodness, which shows itself in a far greater manner still in the institution of the eucharistic bread which nourishes our souls. Oh, how greatly such goodness deserves our praise and our love!

FIRST POINT.

The Goodness of Our Lord in the Multiplication of the Bread which Nourishes the Body.

It was doubtless a great miracle to multiply five loaves of bread and two fishes to such an extent as to satisfy the hunger of five thousand men, and to fill twelve baskets with what remained. All those who were witnesses of this miracle had good reason to proclaim the author of it as King, and to attach themselves to Him so as never to be separated from Him any more. But every day Jesus renews, and will continue until the end of the world, a much more astonishing miracle, namely, the annual multiplication of grains and fruits, sufficient to feed the whole human race, and to give it not only what is necessary, but also what is agreeable ; and the divine action which every year makes all the seeds germinate, grow, and ripen, in such a manner as to provide for all man's wants, in every part of the globe. This striking miracle is hardly ever remarked by ungrateful men. Very few appreciate it, very few thank God for it from the bottom of their hearts. Many even go to the extent of making use of His benefits only in order to offend Him. And yet, O wonder of wonders ! so much ingratitude does not wear out His love. He sheds His dew and His heat upon the field of the sinner as well as on that of the just. Oh, how good God is! what care He takes of His own ! how just it is to love Him, to bless Him, and to offer thanks continually to Him !

SECOND POINT.

The Goodness of Our Lord in the Multiplication of the Eucharistic Bread which Feeds Souls.

There is, in this single fact, a world of miracles. By it Jesus Christ multiplies His presence in as many places as there are altars where the priest sacrifices, in as many hosts as all the ciboriums in the world can hold, in as many particles as each host contains. By it Jesus Christ is always present in these places, remaining, after the sacrifice, in all the tabernacles, although forsaken, solitary, disowned, despised, assaulted by irreverences done to Him, profanations, outrages; and in the midst of it all He prays, He immolates Himself for men, who respond so ill to His love. By it He allows Himself to be distributed as food to all those who present themselves, even to the unworthy, to be borne to all the sick who desire to receive Him, even to the humblest cottage. By it He receives all who desire to speak to Him. He calls to the afflicted in order to console them, the feeble to sustain them, and there is not a moment during the day or night in which He is not happy to grant them an audience. By it He places all His graces at the disposition of those who will receive them, and whoever has recourse to Him may say with Job: "*Deliver me, O Lord, and set me beside Thee, and let any man's hand fight against me*" (Job xvii. 3). Can love go farther than this? And in presence of these miracles, what ought the heart to utter except praises and expressions of love to the God who has so loved men? And what part ought we to take if it be not that of receiving Him often and piously? His desire is to give Himself to us. Shall not our supreme desire be to give ourselves to Him?

Resolutions and spiritual nosegay as above.

Monday in the Fourth Week.

Summary of the Morrow's Meditation.

We will to-morrow resume our meditation upon the motives for contrition, and we shall see : 1st, the ill that venial sin does us ; 2d, the still greater evil that mortal sin occasions us. We will then make the resolution : 1st, to hold in horror our slightest faults, and to keep ourselves in a very humble state before God for having committed any during the course of our life ; 2d, to fly from them as from a pestilence, from the least occasion of sin, to mistrust ourselves, to watch and pray never again to fall into them. We will retain as our spiritual nosegay the words of the publican : "*O God, be merciful to me, a sinner*" (Luke xviii. 13).

Meditation for the Morning.

Let us adore Our Lord Jesus Christ covered with wounds and nailed to the cross for our sins. Our crimes are His executioners, and we are His murderers. O God, my victim, I adore and I love Thee. I deplore my pride, which has crowned Thee with thorns, my effeminacy, which has torn Thy limbs, my love of independence, which has nailed Thee to the cross. O divine crucified

Jesus, form within my heart the hatred of sin, of that evil which is so great that it could be repaired only by Thy death, and make me to understand the evils which it causes myself.

FIRST POINT.

The Evils which Venial Sin causes us.

It would be impossible to express what all the evils are which venial sin causes us. In the next life, if it be not expiated, it will retard the joys of paradise perhaps during long years, and will cost us terrible chastisement. Even after it has been expiated, it will deprive us throughout eternity of the degree of glory and happiness to which an act of the contrary virtue would have elevated us. In this life it cools the friendship of God and diminishes His graces, those graces, nevertheless, so necessary to our weakness; it diminishes our faith and our appreciation of eternal truths; it takes from our soul the tender love of piety, the joys of the Holy Spirit, the delights of innocence; it enfeebles the will, it moulds it little by little to evil, it stifles remorse, dissipates watchfulness, and thence leads to great falls, which are always the consequence of relaxation. Lastly, when it is converted into a habit, it reduces the soul to a state worse than death: to lukewarmness (Apoc. iii. 16). For this frightful state has for its fundamental character the habit of venial faults.

Therefore St. Teresa tells us that God one day caused her to see the place she would have occupied in hell if she had yielded to a temptation of vanity. O my God, how terrible then is venial sin! And yet I fear it so little, I commit it so easily! O my Lord, inspire me with an everlasting horror for it!

SECOND POINT.
The Evils Mortal Sin occasions us.

1st. It takes from us the friendship of God and leaves us a prey to His hatred. Before the fall we were the cherished children of God, His temple, and the objects of His complaisance. We raised to heaven eyes full of confidence, and we saw there a father whose thoughts were full of nothing but love and goodness towards us. But after having committed sin, how great was the change, and what unhappiness was ours! Slaves and the haunt of the devil, children of wrath, and objects of malediction, there is no longer for us anything in heaven, if we are not converted, but a severe judge whose thunder threatens us. Alas! little though we may reflect upon it, how unhappy we are under the weight of the thought, I have incurred the hatred of God! 2d. Sin takes from us peace of heart and leaves us remorse. Whilst we were innocent, we were happy; calm reigned within us, and a sweet and amiable piety reflected outwardly the happiness of a pure heart. But with sin peace disappeared, and gave up its place to distress, to remorse, to anxiety, to the agitation of the conscience, which turns in all directions and everywhere finds nothing but unhappiness. For, O Lord, Thou hast made us for Thee, and outside Thee is neither peace nor happiness. 3d. Sin takes from us all our merits and leaves us naked and indigent. If a man had lived sixty centuries, and at each moment had merited as much as all the saints taken together, a single mortal sin destroys everything, takes from the soul all its merits (Lam. i. 10), and renders it incapable of meriting anything fresh as long as it remains under its empire. 4th. Sin takes from us heaven and leaves us hell. As long as we are in a state of sin, we can no longer aspire either to the beautiful thrones on which we ought to be seated, or to the crowns which ought to encircle our foreheads, or to the enchanting society of the angels and saints, of Mary, of the holy humanity of Jesus Christ, or to the possession of God. The only portion which remains to us is hell. Devils present themselves before God, asking of Him permission to cast the sinners into it (Matt. xxii. 28). What a position, great God! I am only one step removed from hell! Is it not for me that the thunder rumbles? Oh, how foolhardy and imprudent I am! Pardon, my God, mercy!

I deplore my sin, I detest it with my whole soul !

Resolutions and spiritual nosegay as above.

Tuesday in the Fourth Week.

Summary of the Morrow's Meditation.

We will meditate to-morrow upon the firm resolution which forms the essential character of contrition ; and we shall see : 1st, what is the nature and the absolute necessity of it ; 2d, what are its characteristics. We will then make the resolution : 1st, to avoid carefully all occasions of sin ; 2d, not to neglect any means of becoming better, whatever sacrifice it may cost us, or whatever violence we must do ourselves ; and we will retain as our spiritual nosegay the words of the Psalmist : "*I have sworn, and am determined, to keep the judgments of Thy justice*" (Ps. cxviii. 106).

Meditation for the Morning.

Let us adore the Spirit of God inspiring the saints of the Old and the New Testament with a firm resolution, as energetic as it was constant, to lead a perfect life. David exclaims : I have sworn to hate sin ; I hold it in abomination. I have said it ; I am resolved. The right hand of the Most High has worked this change in me (Ps. cxviii.

163 ; Ps. lxxvi. 11). St. Peter allows two inexhaustible fountains of tears to flow from his eyes, and makes amends for his faults by a life devoted to God ; Magdalene changes her profane fires into a furnace of love ; the martyrs take with them to the scaffold a firm resolution not to betray their faith ; St. Ignatius and St. Francis Xavier renounce the world and its glory in order to give themselves wholly to the care of their own salvation. Let us praise the Holy Spirit who caused these great souls to make such sublime resolutions, and let us, with this object in view, render Him all our homage.

FIRST POINT.

The Nature and Necessity of a Firm Resolution.

A firm resolution, very different from those feeble wishes of which hell is full, of those sterile desires which leave us always in the same state, is an energetic determination, a decided resolution to change our life, at no matter what cost ; to be henceforth solidly virtuous, however disagreeable it may be to us ; to do ourselves great violence, and to immolate many of our repugnances. The soul, after having made this firm resolution, does not say, I should be very glad never again to fall, but it says energetically, I will not ; it is my decision ; and if I could begin again, I would rather lose all and suffer all things than to commit the fault of which I have been guilty. It is, lastly, a determination like that which a worldly man makes, not to do such or such a thing which would compromise his fortune, his honor, his liberty, his life. A firm resolution, understood in this manner, is inherent to contrition, and is part of it, because regret for the past necessarily makes the will determine to act in an opposite manner. The motives of the one are essentially the motives of the other; so that without a firm resolution there cannot be true contrition, and consequently neither sacrament nor justification. God will not remit sin excepting in proportion as we are resolved not to fall into it again, and it would be committing a fresh offence against Him to say, I accuse myself and I repent, when in the bottom of our hearts we have an inclination to renew our fault if the occasion should present itself, says Lactantius. Let us here examine ourselves as to how many confessions we have made during our lives without any serious resolution, without any fixed determination to correct ourselves, or else should we be still what we are?

SECOND POINT.

Characteristics and Signs of a Firm Resolution.

A firm resolution ought, like contrition, to be: 1st, universal; that is to say, to extend to all sins, at least to mortal ones, without exception. With God it is all or nothing (James ii. 10); but it ought to apply, above all, to habitual faults; that is to say, to the faults for which the heart has an affection, which makes it fall easily and without great resistance, which even leads it to seek occasions for it. Therein lies the true peril of the soul; the weak side of the place we have to defend against the devil. There, consequently, ought we principally to bring our firm resolution to bear. 2d. A firm resolution ought to be supreme; that is to say, it ought to be superior to all attachments to the extent of breaking them off, to all difficulties up to the point of conquering them, if the service of God requires us to do so. God ought to take the first place; it is His right. 3d. A firm resolution ought to be practical; that is to say, it ought to descend from a general resolution to the means of attaining the end we propose to ourselves. The first means is prayer, that channel of grace without which we can do nothing; the second is vigilance over what we say or do, or what we hear or see, over our thoughts, our intentions, our most frequent faults, above all, our ruling passion; and this vigilance ought to have in view, as its principal object, the separating ourselves from all occasions of sin and of punishing ourselves after each fall. The third means is mortification, which alone can bring back our evil nature beneath the rule of order, enable it to maintain recollection, and make passion die by refusing it that which flatters it. 4th. A firm resolution ought to be persevering. It amounts to nothing to will to do what is right during a certain time; we must will to do it always. He who refuses to God a single moment of his life cannot please Him (St. Prosper), unless he return to Him. Let us examine whether our firm resolution has these four characteristics.

Resolutions and spiritual nosegay as above.

Wednesday in the Fourth Week.

Summary of the Morrow's Meditation.

We will meditate to-morrow upon confession, and we shall see that it ought to be: 1st, humble; 2d, sincere; 3d, entire. We will then make the resolution to impart these three conditions to all our confessions, and we will retain as our spiritual nosegay the counsel of the Holy Spirit: "*Be not ashamed to confess thy sins*" (Eccles. iv. 31).

Meditation for the Morning.

Let us adore Our Lord Jesus Christ instituting, in His love for us, the Sacrament of Penance; let us thank Him for so precious an institution. It is like a sacred bath which washes away all our stains, like a divine channel through which grace flows upon us; it is a school of wise counsels and encouragements to good; it is, lastly, the most efficacious means for correcting our defects and making us advance in the practice of virtues. May we always rightly use this marvellous invention of divine love.

FIRST POINT.
Our Confessions ought to be Humble.

We ought to present ourselves before the priest filled with respect and confusion, like a sinner before an angel of God, before another Jesus Christ; like a sick man covered with hideous wounds before a doctor who can cure us if we show them to him such as they are; like a criminal who has committed high treason against the Divine Majesty, who deserves to be cast into the dungeons of hell, before the supreme Judge who holds in His hands the sentence giving us life or condemning us to eternal death. Not being able to obtain anything in the name of justice, but solely in the name of clemency and mercy, we ought to present ourselves with profound exterior and interior humility, humbly confessing our sins, and declaring them, not with the indifference of a person reciting a history, but with the shame and grief of a soul which understands its wrong-doing; not in accusing ourselves in order to avoid the confusion of appearing guilty, but in accusing ourselves without the contrivances which tend to induce the belief that our sins are less than they really are in the sight of God; not with pride and arrogance, as if we had performed some beautiful action, but with modesty and sighs over our wretchedness and as fearing the judgment of God more than that of man. Is it thus that we make our confessions?

SECOND POINT.

Our Confessions ought to be Sincere.

Sincerity in confession consists in confessing with entire candor and simplicity all we remember, without troubling ourselves about what we may possibly forget, since a defect of memory is not a sin in the sight of God. It is not right to exaggerate our sins, under the pretext that it is better to say more than less; the sick man is not wise who exaggerates what he suffers when he speaks to his doctor. It is still worse to veil our faults by artificially enveloping them with other accusations which are less painful, and gliding rapidly over what costs us the most, in the hope that the confessor will not observe it. A sincere penitent desires only to be known for what he is, and he holds cunning and artifice in horror. It is equally wrong to excuse our faults, whilst at the same time saying what they are, or to endeavor to make them seem less by laying the blame of them on others, as Adam and Eve did, for to do so is a want of candor. But the supreme degree of evil is to hide our faults through false shame. Then the sacrament of mercy is changed into an anathema, the work of salvation into a work of reprobation, and the sentence of life into a sentence of death. Better a thousand times not to confess. We may deceive man, but we cannot deceive God, who knows the secrets of all hearts; for one sin hidden from the priest all those which we have committed will, at the judgment day, be brought to light in the face of the whole universe; and for a little shame that we try to avoid in this life we shall be covered in the next with eternal confusion. Let us here examine our confessions. Have we declared them without disguise, without excuse, without ingeniously presenting them in colors which disguise their deformity? Have we confessed certain sins as certain, doubtful ones as doubtful, and have we avoided superfluous and useless words, terms which are vague, obscure, and equivocal, and which prevent our confessor from rightly seeing the truth?

THIRD POINT.

Our Confessions ought to be Entire.

In order that they may have the necessary integrity, it is not enough to accuse ourselves of mortal sins, we must also, 1st, say how many times we have fallen into them, declare the aggravating circumstances of them, or such as change their character, the results or grievous consequences of them—for example, if there has been scandal, if the calumny has been serious in its nature, uttered before several persons, against a superior, or a priest; if it has been inspired by hatred, by resentment, or vengeance, and, when we accuse ourselves of disobedience, whether the disobedience was accompanied by arrogance, contempt, or bad temper. Without that, the confessor is not sufficiently acquainted with the state of his penitent to pronounce a prudent judgment upon him. We must, 2d, accuse ourselves of venial faults. Although this is not a rigorous precept, it is at the same time very important to do it: 1st, because not to accuse ourselves of a sin of which we have a doubt whether it be mortal or venial would be sacrilege, and there is often a reason for such a doubt; 2d, because the confessor, knowing the penitent only imperfectly, cannot direct him with surety either as to the communions to be permitted to him, or in regard to the other acts of Christian life, or in the reform of his faults, or the acquisition of virtues; 3d, because the accusation of venial sins makes the penitent take greater care to avoid them, and he is helped therein by the grace of the sacrament, by the advice of the confessor, and the shame of the accusation. Let us examine ourselves as to whether such have been our confessions.

Resolutions and spiritual nosegay as above.

Thursday in the Fourth Week.

Summary of the Morrow's Meditation.

We will meditate to-morrow upon the third part of the Sacrament of Penance, which is satisfaction, and we shall see: 1st, its importance; 2d, its extent; 3d, the manner in which to acquit ourselves of it. We will then make the resolution: 1st, to perform our penance always as soon as possible after confession, accompanying it with a great desire to become better; 2d, cheerfully to suffer all the crosses which Providence sends us, and to add to it some voluntary mortifications,—for example, in our meals, our curiosity or the desire of our eyes, our personal comforts, etc. Our spiritual nosegay shall be the words of the Council of Trent: "*The whole of the Christian life ought to be a perpetual penance.*"

Meditation for the Morning.

Let us adore Jesus Christ satisfying for our sins, and with this object in view, embracing a life of suffering. He is born in extreme poverty, He lives a life of continual labor, and He dies in the most cruel torments. O admirable model of penance! beautiful example for those who, animated by zeal for the justice of God against themselves, desire fully to satisfy Him for their sins!

FIRST POINT.

The Importance of Satisfaction, even after the Sin is Pardoned.

Every sin deserves a penalty, every injury a reparation. Our sin deserved eternal punishment; by the Sacrament of Penance God changes it into a temporal penalty. Thou dost pardon, Lord, the sinner who confesses his sin, says St. Augustine, but on condition that he will punish himself for it. And what can be more just? Is it right that the Innocent, that the Man-God should suffer the most cruel of deaths for sin, and that the guilty should receive the price of His death without taking part in the expiation? St. Paul did not think so when he said, "*Fill up those things that are wanting of the sufferings of Christ*" (Coloss. i. 24). "*If we suffer with Him we may be also glorified with Him*" (Rom. viii. 17). The Church is of the same opinion when she calls penance a laborious baptism, which justifies the soul only on the condition that many tears are shed and much penance endured (Conc. Trid. Sess. xiv. c. xi.). Thus God pardons in His goodness, but in His justice He requires a satisfaction. Satisfaction performed by man is incapable, through its own merit, of satisfying His justice, but in His goodness He authorizes man to take advantage of the satisfactory works of the Saviour, by uniting them with His, and thereby appropriating to Himself their infinite value. In this manner justice and goodness are fully satisfied. Let us admire and bless this marvellous arrangement of divine wisdom.

SECOND POINT.

The Extent of the Satisfaction Due to Pardoned Sin.

If the penance imposed by the confessor is, generally speaking, very slight, it is only from a fear of discouraging the penitent by requiring more; but in reality a very different satisfaction is due. We owe to God, says Tertullian, a penance which shall be a compensation, and, as it were, an abridgment of everlasting fires. And the Council of Trent adds that the whole of the Christian life ought to be a perpetual penance. If God pardoned Adam and David, it was only on condition that they should be punished with dreadful penalties: the one in himself and in all his posterity; the other in his person and in his people. The saints, after having received pardon, did not on that account the less devote themselves during their whole life to austere penance; lastly, the just in purgatory, although God pardons them, are not the less obliged to suffer torments compared with which all the sufferings of life are slight. O justice of God, how severe thou art, and what enemies we are of ourselves in performing so little penance in this world!

THIRD POINT.
The Manner of Satisfying God for our Sins.

1st. We must scrupulously perform the penance imposed by the confessor. Particular graces are attached to it, inasmuch as it is an integral part of the sacrament; and, on the other side, to omit it would be to mutilate the sacrament and thereby wound Jesus Christ. To retard it would be to diminish the merit of it, which would enable us to lead a better life; it would be to diminish the merit in question by the venial sins which we should commit in the interval; it would even be to lose it entirely if, during the interval, we were to fall into mortal sin. Lastly, it would be to fail as regards the object of the penance, since it is often given us as a preservation against a relapse, or a remedy for our sin, or as a means of sanctifying certain feast days. 2d. We must receive this penance with respect and submission, as being imposed upon us by Jesus Christ in the person of His minister; looking upon it as being infinitely below what our sins deserve, and accomplishing it devoutly with a great desire to lead a better life, and with sorrow for the past. 3d. To this sacramental penance we must unite the endurance of all the difficulties of our position or state of life, all the infirmities of our bodies, the trying temperature of the seasons, the various disagreeable circumstances of life, the defects of our neighbor, accepting all these crosses in a spirit of penance, and often saying to ourselves: What is this in comparison with hell, where I have deserved to burn forever and ever? 4th. We must, in the same spirit of penance, refuse to listen to the demands made upon us by our effeminacy and sensuality; we must renounce dangerous amusements, and such as are useless or too much prolonged, the satisfaction of curiosity, of self-will, of self-love, of our temperament; we must find our pleasure in doing our duty and deprive ourselves of the rest, saying with that holy penitent who was asked to join in parties of pleasure, of the table, and of the play: "I leave all that to righteous souls; as for me who have sinned and am in danger of sinning again, my portion is to sigh and to perform penance" (St. Pacian, ap. Baron.). Do we put in practice these various ways of satisfying the divine justice?

Resolutions and spiritual nosegay as above.

Friday in the Fourth Week.

Summary of the Morrow's Meditation.

We will meditate to-morrow upon the Feast of the Precious Blood, which the Roman Church celebrates to-morrow, and we shall see : 1st, the gratitude due to Jesus Christ for the gift which He has made us of His blood ; 2d, the practical consequences which result from this ineffable gift. We will then make the resolution : 1st, to love more and more Jesus Christ, who has so loved us, and to serve Him with more generosity than ever ; 2d, to place all our confidence in the merits of His blood, and never allow ourselves to yield to discouragement or mistrust. Our spiritual nosegay shall be the words of St. John, "*Jesus Christ hath washed us from our sins in His own blood*" (Apoc. i. 5).

Meditation for the Morning.

Let us adore Jesus Christ making us a gift of the whole of His blood, down to the last drop ; let us thank Him for this ineffable gift ; let us love Him for so much love, and ask of Him grace to profit by it.

FIRST POINT.

Gratitude due to Jesus Christ for the Gift which He makes us of His Blood.

If a man were to give another man the whole of his fortune, it would be, doubtless, a great gift, above all if it were considerable. What would it be, then, if he were to give him his blood, and shed the whole of it for him? It would evidently be love carried to its most extreme limit. Now, this is what Jesus Christ has done in regard to us; and here let us observe, 1st, the value of His blood. It infinitely surpasses the value of any human blood, because it is the blood of a God, by virtue of the hypostatic union, blood, consequently, of an infinite value. This blood is offered by a God in every sacrifice offered to the Divine Majesty; and the dignity of a God-Priest offering the blood of a God-Victim communicates to it a fresh degree of infinite value. Let us observe, 2d, the marvellous effects of Christ's blood. It extinguishes the fire of divine justice, irritated by our crimes (Rom. v. 9). It is the host of expiation for our sins (Rom. iii. 25). It is the price of our redemption (Eph. i. 7). It is the bath which purifies our conscience (Heb. ix. 14). It is the seal of peace between heaven and earth (Coloss. i. 20). It opens heaven and closes hell under our feet (Heb. ix. 12). Far from crying out for vengeance, like the blood of Abel, each drop of this blood cries out for mercy (Heb. xii. 24). Let us remark, 3d, that this blood, so great as it is in value, is given to us not with an avaricious hand, but with incomparable generosity. Although a single drop of it would have sufficed to wash away the sins of a thousand worlds, Jesus Christ gives the whole of it, and He even gives it to those whom He foresees will be but little worthy of it; He gives it, not once only, but millions of times. He begins to shed it eight days after His birth, beneath the knife of circumcision; He sheds it in the Garden of Olives, where a sweat of blood inundates the ground; He sheds it at the flagellation, at the crowning with thorns, at the crucifixion, at the opening of the sacred side; He offers it every day at the holy sacrifice over the whole surface of the globe, and gives it to us to drink in the Holy Communion; He keeps it in all the tabernacles of the world; and in them His blood ceaselessly begs for grace for us. Lastly, He applies the merits of it to us in the sacraments, which are as so many channels by which this adorable blood is communicated to souls. What gratitude, then, do we not owe the Saviour for this prodigality of His blood towards poor sinners like ourselves!

SECOND POINT.

Practical Consequences from these Considerations.

1st. There must be great generosity in the service of Jesus Christ. When a God gives His blood, are we not inexcusable if we refuse Him the sacrifice of our will, our actions, and our pleasures? When we possess in our bosom the blood of Jesus Christ, when we are of such noble and divine blood, we must be generous in spirit and must be inspired by the noble sentiments to which nothing is so dear as sacrifice. 2d. We must honor this precious blood by assisting devoutly and frequently at the holy sacrifice, by often approaching the sacraments, by correspondence to the interior and exterior graces which are the fruits of it; by the often reiterated offering of our actions and of our heart in a spirit of gratitude. 3d. We ought to have an unlimited confidence in the merits of the divine blood. Let those be troubled and be wanting in confidence who do not know the price of the blood of the Saviour; but when we know by faith that Jesus Christ has left at our disposition all the merits of His blood, with the faculty to apply them to ourselves by prayer, by the sacraments, and by the holy sacrifice, we are inexcusable if we are wanting in confidence. With our crucifix in our hands our courage ought never to fail. It is true, O Jesus, I cannot say, "*I am innocent of the blood of this just Man*" (Matt. xxvii. 24), since it is my sin that has delivered up this innocent blood, but I will say, animated by another spirit than that which animated the Jews: May His blood fall upon me (Matt. xxvii. 25) to wash away my iniquities and preserve me from the exterminating angel, like the blood of the paschal lamb on the doors of the houses of the ancient people. Am I faithful in deriving these fruits from the Passion of the Saviour?

Resolutions and spiritual nosegay as above.

Saturday in the Fourth Week.

Summary of the Morrow's Meditation.

We will meditate to-morrow as a supplement to our meditations upon the Sacrament of Penance: 1st, on the obligation of allowing ourselves to be guided by our confessor; 2d, on the manner in which this direction ought to consist. We will then make the resolution: 1st, to take counsel with our confessor respecting our rule of life and the employment of our time, the reformation of our defects, the practice of virtues, and the kind of good works for which we are best suited, if we are in a position to perform them; 2d, to consult our confessor in the difficulties and doubts we may meet with. We will retain as our spiritual nosegay the words of the Holy Spirit: "*Seek counsel always of a wise man*" (Tob. iv. 19).

Meditation for the Morning.

Let us adore Our Lord Jesus Christ as regards the manner in which He guided St. Paul after his conversion; that great apostle aspired to nothing else excepting to know and fulfil the will of God (Acts ix. 6). Our Lord, instead of enlightening him Himself, or leaving him to his own guidance, in the state of supernatural light which surrounded him, sends him to a wise director (Ibid. 7). Let us thank Him for this beautiful lesson, which teaches us not to lean upon our own prudence (Prov. iii. 5), and always to take the advice of a wise man (Tob. iv. 19).

FIRST POINT.

The Obligation of Allowing Ourselves to be Guided by our Confessor.

"No one is sufficient to himself so that he can lead himself" (St. Basil, *Orat. de Felic.*). Our reason deceives us; the wisest lose themselves when, instead of taking counsel, they trust in their own lights, says St. Bernard (*Ep.*, lxxxii.). He who sees in his confessor only a confidant of his sins in order to receive absolution of them, and not a counsellor to direct him in the road of life, is as much exposed to lose his soul as is a ship without a pilot, a blind man without a guide, a sick man without a doctor; and the devil knows no surer way of making Christians lose their souls than by inspiring them with the presumptuous opinion that they can govern themselves by their own sole judgment (St. Dorotheus, *Doct.* 5). Therefore all the saints have been faithful to the practice of taking counsel respecting their own conduct (St. Vincent Ferrer, *De Vita Spirit.*). Moses takes counsel of the ancients; David is reproved by Nathan and Gad, who are not such great prophets as he is; Saul is sent to Ananias by Jesus Christ, who could have instructed him; lastly, the Saviour Himself listened to and questioned simple men (Luke ii. 46). It is in the order of Providence that men should be instructed by other men and should depend upon one another for guidance; it is also in the order of reason: he who sees clearly into the conscience of another does not see clearly into his own, he deludes himself in respect to his obligations, his vices and his virtues, his merits and his aptitudes; and all have need of a wise counsellor who studies them without prejudice and with the grace of his ministry. It was this which made Bourdaloue, when he was preaching in Paris, utter these remarkable words: "I cannot sufficiently deplore the blindness of people living in the world who desire to have confessors and not directors, as though the one could be separated from the other " (Sermon for the thirteenth Sunday after Pentecost). Let us here examine ourselves. Are we not of the number of those who would have made the holy preacher sigh?

SECOND POINT.
The Manner in which the Direction ought to be Performed.

1st. We must see in our director, not a man or a sage, but an angel, clothed with the wisdom of God, a Jesus Christ, nay, even a God, like the holy solitary who said (John Clim., *Grad.* 4) : "I see the image of Jesus Christ in my superior." We must speak to him, consequently, with entire openness of heart, and with perfect confidence, as to a charitable physician and the faithful friend whom God has given us for our guidance; reveal to him all the good and all the evil of which we are aware in ourselves, our inclinations, our intentions, our temptations, without reserve, without disguise, without any of those artifices of which self-love sometimes makes use in order to make the will of our director harmonize with our desires; we ought to put on one side all human respect, all shame, all repugnance, as well as all feelings of vanity or curiosity. 2d. We must listen to his counsels with respect and confidence, and follow them with fidelity and exactness, however contrary they may be to our own judgment, our character, and our will. 3d. We must abandon ourselves so entirely to his guidance in all things that have relation to our salvation, that we shall never undertake anything without first consulting him about it ; that we never refuse to do what he tells us, that we give him an absolute and entire liberty to tell us what he thinks ; that we never discuss his counsels, but embrace them as being the best ; if we have doubts, that we expose them with as much indifference as freedom, and that, whether he says one thing or another, we are equally obedient to him. Are these our dispositions and our manner of acting?

Resolutions and spiritual nosegay as above.

Passion Sunday.

The Gospel according to St. John, viii. 46–59.

"At that time Jesus said to the multitudes of the Jews : Which of you shall convince Me of sin? If I say the truth to you, why do you not believe Me? He that is of God heareth the words of God. Therefore you hear them not, because you are not of God. The Jews therefore answered, and said to Him : Do we not say well that Thou art a Samaritan and hast a devil? Jesus answered: I have not a devil; but I honor My Father, and you have dishonored Me. But I seek not My own glory; there is One that seeketh and judgeth. Amen, amen, I say to you, if any man keep My word he shall not see death forever. The Jews therefore said: Now we know that Thou hast a devil. Abraham is dead and the prophets: and Thou sayest, if any man keep My word he shall not taste death forever. Art Thou greater than our father Abraham, who is dead? and the prophets are dead. Whom dost Thou make Thyself? Jesus answered: If I glorify Myself, My glory is nothing. It is My Father that glorifieth Me, of whom you say that He is your God. And you have not known Him; but I know Him. And if I shall say that I know Him not, I shall be like to you, a liar. But I do know Him, and do keep His word. Abraham your father rejoiced that he might see My day: he saw it, and was glad. The Jews therefore said to Him: Thou art not yet fifty years old, and hast Thou seen Abraham? Jesus said to them: Amen, amen, I say to you, before Abraham was made, I am. They took up stones therefore to cast at Him: but Jesus hid Himself, and went out of the temple."

Summary of the Morrow's Meditation.

The holy fortnight on which we are about to

enter is destined for the honoring of the wounds of Our Saviour; and that we may appreciate how much love Jesus shows us in this mystery we will consider: 1st, who He is that suffers, and for whom He suffers; 2d, how much He suffers; 3d, what benefits He obtains for us by His sufferings. Our resolution shall be: 1st, to pass this fortnight in special sentiments of piety, of recollection, and of love towards the crucified Jesus; 2d, to keep the crucifix habitually before our eyes, and often and lovingly to kiss it. Our spiritual nosegay shall be the words of the Apostle: "*He loved me and delivered Himself for me*" (Gal. ii. 20).

Meditation for the Morning.

Let us adore Jesus Christ overwhelmed by a sea of suffering and of ignominy. Let us say with the Apostle: It is His love for me which has reduced Him to this state. Let us love and bless so much love; let us compassionate so much suffering.

FIRST POINT.

Who is He that Suffers? Who is he for whom He Suffers?

Nothing is better suited than the contrast between these two thoughts to show what is the love of Jesus Christ for us in the midst of His sufferings. Great God! exclaims St. Thomas, even if

Thou wert my slave, and I Thy master, there would be in the devotedness of a slave who should suffer such great things for his master an heroic love capable of overwhelming with amazement the most insensible of souls. What ought I, then, to think of the contrary supposition, which is the only true one? For we know well that it is the God of Calvary, it is the Lord and Master of all things, who humbles Himself and dies for His servant; it is the Eternal King of Ages who immolates Himself for His servant; it is God dying for a worm. O abyss of love! Again, if he for whom this God humbles Himself and dies were a friend worthy of His interest and of His love—but no; he for whom this God dies is at one and the same time nothingness by nature, since he has only a borrowed existence, and is sin by his origin and sin by malice. It is baseness itself in revolt against God; and God, against whom he has rebelled, dies to expiate his rebellion! It is a frightfully audacious creature who has dared to offend his Creator, and the offended God wills to die for his having offended Him! It is an ungrateful wretch who will not feel, and God knows it well, any gratitude for such great devotion; who will look with a dry eye and an insensible heart at the figure on the cross; who will coldly celebrate the holydays consecrated to the memory of so touching a mystery; and, what is still worse, it is a traitor who will violate his oaths, who will recommence his insults, who will crucify his God afresh in so far as it depends upon him, and that not once, but thousands of times; and yet, for a creature so abominable, so worthy of the anathemas of heaven and of earth, a God humbles Himself and dies! O abyss of love! O fathomless mystery of love!

SECOND POINT.

The Greatness of the Sufferings of the Saviour.

Here new abysses of love open themselves up. Jesus Christ might, with a single drop of His blood, a single tear from His eyes, have redeemed the whole human race; but as love is more shown in proportion to the greater suffering endured for its sake, He gave Himself up to all kinds of suffering and of ignominies. He sacrifices all; first, His liberty, for He allows Himself to be bound like a captive; then His honor, for He consents to pass for a fool, for a criminal, for a blasphemer, for a man worse even than Barabbas, who was a thief and an assassin—worse than the two thieves between whom He was crucified, as being the most guilty of the three. He sacrifices His body; for, from the sole of His foot to the top of His head there are nothing but open wounds, blood which flows and bones exposed. He sacrifices His soul; for He gives it up as a prey to the anguish of death (Matt. xxvi. 38), to the abandonment of creatures and of His own Father (Ibid. xxvii. 46). Lastly, He sacrifices His life; for love immolates Him upon the altar of the cross (Is. liii. 7); and, with His own free consent and His perfectly free will, He offers Himself for us to His Father (John x. 17, 18). O love! how incomprehensible thou art! How deep are thy abysses! And we—how have we responded to so much love? What have we done for Him who has done so much for us?

THIRD POINT.
The Immense Benefits Procured for us by the Sufferings of the Saviour.

The generosity of a benefactor is measured not only by the greatness of the sacrifices which he makes, but also by the excellence of the benefits which he bestows; and here are displayed fresh abysses of love! For the benefits which the Passion of the Saviour procures for us are really ineffable. They are, 1st, heaven opened and hell closed, death and sin vanquished. Without the Redemption, the whole human race was damned; by the Redemption, he who wills it can be saved, and those alone are damned who will to be damned. They are, 2d, the titles of children of God, of heirs of the eternal Kingdom, of co-heirs and of members of Jesus Christ. What happiness and what glory! They are, 3d, faith, without which we should be like the heathen nations—without belief and without morals; hope, which consoles and supports; charity, which unites men to each other and to God; the Church, which teaches and directs us; the priesthood, that sun of the moral world; the holy sacrifice of the Mass, that mysterious link between heaven and earth; the sacraments, those channels through which the blood of the Saviour flows, bearing everywhere grace, strength, and life. Happy the fault of Adam, which was the means of our having such a Redeemer! (Bened. cer. pasch.) But woe to us if we abuse so many graces! Let us now at last decide better to love and better to serve the Author of these benefits.

Resolutions and spiritual nosegay as above.

Monday in Passion Week.

Summary of the Morrow's Meditation.

After having meditated how greatly the crucified Jesus has loved us, we will now meditate upon how we ought to love Him ourselves, and we shall see that we ought to love Him, 1st, with a penitent love in memory of the past; 2d, with a generous and fervent love for the present and the future. We will then make the resolution:

1st, to address frequently, during the day, loving aspirations of love to Jesus suffering and dying for us; 2d, to perform all our actions from a motive of love for Him, and to give, with this object in view, all the perfection of which we are capable to these actions. Our spiritual nosegay shall be the words of St. Paul: *"Christ died for all, that they who live may not now live to themselves, but unto Him who died for them and rose again"* (II. Cor. v. 15).

Meditation for the Morning.

Let us prostrate ourselves in spirit at the feet of Jesus Christ suffering and dying for us, and let us offer Him our most fervent homage of adoration, of gratitude, and of love.

FIRST POINT.
We ought to Love Jesus Crucified with a Penitent Love in Memory of the Past.

How full of shame for us, what a subject for regret and repentance, is our whole past, studied at the foot of the cross! Alas! is it not true that the cross of the Saviour has most often found in us nothing but lukewarmness and insensibility, perhaps even coldness and baseness? Is it not true that the cross is like a great book, in which our sins are written in characters of blood? The flesh of the divine Saviour, which is torn to pieces, and His blood, which flows under the lashes of the scourges, are an accusation against the unruly love we have for our body. His head, crowned with thorns, reproaches the pride of our minds and the vanity of our thoughts. The gall and vinegar which He is given to drink protest against the effeminacy and sensuality of our tastes. His face, wounded with blows and covered with spittle, condemns our desire to make a parade and attract notice, our horror of humiliation and contempt. The nails which fasten Him to the cross ought to make us blush for our love of liberty and of our inborn independence. Lastly, His death speaks to us of the enormity of our sins, which are the cause of it. O Jesus, whom I ought so much to love, how I regret to have so greatly offended Thee! Penitence ought to be my portion forever; and, instructed by the voice which issues from all Thy wounds, I will begin a new life.

SECOND POINT.

We ought to Love Jesus Crucified with a Generous and Fervent Love.

If a man were to show us kindness, we should not be insensible to it. If he were to sacrifice for us his fortune, we should think that we could never thank and love him sufficiently. What would it be, then, if to the sacrifice of his fortune he were to sacrifice his honor and sacrifice his liberty to the extent of allowing himself to be bound and scourged like a slave? What would it be, above all, if he were to sacrifice his life in order to save ours? Can we conceive a heart sufficiently bad to offend such a benefactor, or to refuse him a sacrifice, no matter what it might be? O crucified Jesus, who hast done all this and infinitely more still—who hast heaped ineffable benefits upon us, which were the cause of Thy holy death, how then can we have the heart to offend Thee? to refuse Thee aught, when Thou givest all, when Thou givest Thy own self without reserve? How can we be attached to earthly possessions when Thou art all naked upon the cross? How can we indulge in self-love and vanity when Thou art covered with confusion? How can we give way to self-will when Thou dost obey even unto death? to pleasure and enjoyment, when for us Thou didst taste suffering? No, my God, it is not possible. To Thee is due a generous love which spares nothing, which sacrifices everything without reserve. But even that is not enough. To this generous love ought to be united fervor; that is to say, that noble and delicate sentiment, which, after having given all, humbly confesses that it is a million times too little; that it is nothing in comparison with what Thou dost deserve, O crucified Jesus! Such was the love of the saints. They always aspired to love more and more, and, whatever they did, to do a thousand times more, and a thousand times more still. They consumed themselves with holy desires to love always more. They would have desired to love infinitely if they had been able, because they comprehended that our great God is millions of times worthy of an infinite love. Hence it was that on one side they never relaxed their efforts, and always made progress; and on the other were always very humble, ashamed and confused not to love more. Oh, who will give us this fervent love which burns ceaselessly like a living flame and is fed in consuming itself? O love, come to me, consume me; may I no longer live except by love, and may I die of love! O crucified Jesus, give me, like St. Paul, a heart able to say: The love of Jesus Christ constrains my heart, and nothing can stay its holy ardor (II. Cor. v. 14; Rom. viii. 37).

Resolutions and spiritual nosegay as above.

Tuesday in Passion Week.

Summary of the Morrow's Meditation.

We will meditate to-morrow upon how we ought to love the cross: 1st, because it is our salvation; 2d, because it is our consolation in the troubles of life. We will then make the resolution: 1st, to keep ourselves habitually in spirit at the foot of the cross during these holy days, and often to press our lips to it; 2d, to have recourse to the cross in all our trials. Our spiritual nosegay shall be the words of St. Paul: "*With Christ I am nailed to the cross*" (Gal. ii. 19).

Meditation for the Morning.

Let us prostrate ourselves at the feet of Jesus on the cross; let us lovingly kiss His sacred feet. It is there that the Christian abundantly finds salvation for eternity and consolation in the present life; that is to say, happiness in heaven and happiness upon earth. To Jesus crucified be adoration, love, thanksgiving, and benediction.

FIRST POINT.

We ought to Love the Cross because it is our Salvation.

There are two kinds of crosses—the cross of Jesus Christ, upon which He died, and our personal crosses, which are our daily trials. Now, these two kinds of crosses merit all our love, because both the one and the other are the cause and the instrument of our salvation.

1st. The cross of Jesus Christ, because without it, children as we were of wrath and slaves to the devil by our birth, we were lost everlastingly, and by it Jesus Christ cast down the infernal powers, tore away from their hands, says St. Paul (Coloss. ii. 14), the sentence which condemned us, effaced it with His blood, and nailed it to the cross, in order that no hand might take it away. He chained to His cross, as to a triumphal car, the inimical powers; He despoiled them and led them away captive, so that now every one may be saved who desires to be saved. The cross makes to flow throughout the whole Church, by means of the sacraments, by the holy sacrifice of the Mass, by holy thoughts and pious emotions, all the graces of which it is the source and the inexhaustible ocean; it offers to all pardon for the past, courage for the present, confidence for the future. Is not this enough to merit all our love?

2d. We ought to love our personal crosses, because the cross of Jesus Christ has raised them to the distinguished honor of being the most efficacious means of perfection, and the warrant of our eternal hopes. Patience, which endures the cross, says St. James, is perfection, and solid perfection, because it has been proved in the crucible (James i. 4). It is, according to St. Paul, the crown of faith (Philipp. i. 29). It is the warrant and the joy of hope. For a moment of light suffering, an immense weight of glory (II. Cor. iv. 17); after trial, the crown of life (James i. 12). It is one of the beatitudes proclaimed by Jesus Christ: *"Blessed are they that suffer"* (Matt. v. 10). It is a special grace which God sends to His best friends; it places them on the royal road to heaven. It suffices to have only a little faith in the words of the Saviour to esteem a good cross more than all riches; a good affront borne in a Christian manner more than all honors; humiliations, even the most mortifying, more than all crowns; ignominy more than all applause; confusion more than all kinds of praise. Therefore the Gospel says: Receive crosses not only with patience, but with gladness (Matt. v. 12). And St. James adds: Receive them with every kind of joy, that is to say, with the joy of the poor who receive immense riches, with the joy of the man chosen amongst the people to receive a crown, with the joy of the laborer who gathers together a rich harvest, with the joy of the merchant who amasses a great gain, the joy of the general who obtains a great victory (James i. 2). Thus, also, thought the saints; St. Paul, when he said, I abound with joy in all my tribulations (II. Cor. vii. 4), and St. Andrew, when at the sight of the cross he cried out lovingly, "Oh, welcome, good cross, welcome, and ever longingly desired!" Are these our sentiments?

SECOND POINT.

We ought to Love the Cross because it is our Consolation in the Troubles of Life.

A heathen guessed this truth when he said that *by accepting trials cheerfully they are softened* (Horatius). Before him the Holy Spirit had said: "*Whatsoever shall befall the just man, it shall not make him sad*" (Prov. xii. 21). What is it then under the New Law, where Jesus Christ crucified presents Himself to the afflicted soul, to say to it, poor soul, be consoled, I pity thy trials; I know what suffering costs thy nature; I have passed like thee through trials; and if, to console thee, thou requirest a friend who understands suffering, I possess in a supreme degree the character of a true consoler. In past times a great monarch and his minister, taken in war, were stretched by a cruel conqueror upon burning braziers. The minister uttered loud cries, and I, said the monarch to him, am I on a bed of roses? I can hold the same language to thee, O afflicted soul! Behold My head, crowned with thorns, My whole body torn, My whole person a victim to ignominy; I have suffered all this from love for thee; wilt thou not be willing to suffer infinitely less from love for Me? When I drank the chalice down to the dregs, wilt thou refuse to taste at least only a few drops? Courage, have patience; thou shalt reign one day with me; come to the throne by the same path. Unite thyself with me who am thy God and suffer from love for Me (Ecclus. ii. 3). Thanks, O my God, for this precious balm with which Thou anointest my wounds. Ah, Thou art indeed the Consoler of the afflicted soul. O holy crucifix! I take you in both my hands! I press you to my heart and to my lips, and I feel myself consoled!

Resolutions and spiritual nosegay as above.

Wednesday in Passion Week.

Summary of the Morrow's Meditation.

We will consider to-morrow that we ought to love the cross, because we find in it: 1st, our strength; 2d, our glory. Our resolution shall be: 1st, to remember the cross in our seasons of weakness or discouragement, in order to revive our courage; 2d, no longer to have any care for the vain glory of the world, and to attach ourselves solely to the solid glory of the cross. Our spiritual nosegay shall be the words of St. Paul: "*God forbid that I should glory save in the cross of Our Lord Jesus Christ*" (Gal. vi. 14).

Meditation for the Morning.

Let us prostrate ourselves in spirit before the cross of the Saviour, and let us offer to it the homage of our most fervent piety, of our adoration, love, and praise (Ps. xciv. 6).

FIRST POINT.
We ought to Love the Cross because it is our Strength.

Man is feeble of himself, and, on the other side, he is in such critical positions, he has faults and passions which are so difficult to overcome, virtues which are so painful to practise, that it is necessary that a supernatural strength should come to the succor of human weakness. Now, it is in the cross that this strength is to be found. We find in it an example which puts to shame our pusillanimity and excites our courage; a guarantee of our immortal hopes which, raising our heart to heaven, renders it stronger than the whole earth; a grace which sustains, a love which provokes our love and inspires devotedness; lastly, the seal of the elect, which invites us to walk in the same path as they did in order to reach the goal which they have attained. St. Paul attached himself to the cross (Gal. ii. 19), and leaning upon it he esteemed himself to be stronger than all kinds of temptations and trials (Rom. viii. 37). The martyrs and the confessors in their torments thought of the cross, and found therein a strength which rendered them invincible. I suffer greatly, said one of them, but what is it compared with what Jesus suffered on the cross? Let us imitate these beautiful examples. Are we tried by reverses of fortune, even to the extent of suffering the most extreme poverty, the nakedness of Jesus on the cross will render privation dear to us, and will make us exclaim courageously with St. Jerome: I will follow naked Jesus Christ naked. Are we afflicted in our body by infirmity and suffering, the wounds of Jesus Christ on the cross will make us cherish suffering, and enable us to say with St. Bonaventura: *"I will not live without suffering when I see Thee suffering;"* or with St. Teresa: *"Either suffering or death!"* I have a horror, said this great saint, of enjoyment and comfort, of sensuality and effeminacy. Are we a butt for calumny, to want of consideration, to contempt; the opprobrium suffered by Jesus on the cross will destroy our illusions in respect to the love of esteem and of praise. We henceforth have no longer any desire for them; for how can we have any respect for the esteem of a world which has so ill appreciated eternal wisdom? How can we desire to be treated better and to be more honored than a God? Lastly, have we interior troubles to suffer, a character to reform, self-will to overcome, the meekness and obedience of Jesus on the cross will render us meek and docile, simple and obedient. Thus, in whatever position we may be, whatever may be the troubles in us or around us, the cross will be our strength; with it we shall triumph over all difficulties, with it we shall be happy in the midst of suffering, rich in poverty, content amidst contradictions.

SECOND POINT.

We ought to Love the Cross because it is our Glory.

The cross and sufferings are so great an honor that our sins deserve that we should be deprived of them, and that we should be condemned to the riches, the honors, and the pleasures of the world, against which Our Lord pronounced this terrible anathema, "*Woe to you that are rich, for you have your consolation*" (Luke vi. 24). The soul on which God bestows these false positions ought to be humiliated and confounded, and ought to fear condemnation at the day of judgment. The soul, on the contrary, that is favored by God with the gift of the cross ought to be afraid of indulging in pride, because then it is treated like a God, assimilated to Jesus Christ, the true God, and, like Him, fed with sufferings, opprobrium, and poverty. The world, which entertains false ideas respecting glory, does not at all understand this language; nevertheless what is there which is clearer? According to the world, glory consists in the nobility of an illustrious blood; but the cross gives to the Christian a nobility higher than all earthly nobility; **by** means of it the Christian is a child of God, with a right to say to God: Our Father, who art in heaven; he is the brother of Jesus Christ and co-heir of the heavenly kingdom. According to the world, glory consists in the possession of vast domains; but the cross assures to me heaven for my inheritance, a throne on which I shall judge the world (Eph. ii. 6), and infinite benefits compared with which the whole world is as nothing. According to the world, glory consists in the superiority of mind by which so many sages of past days were distinguished; but in comparison with the hidden mystery of the cross all the wisdom of the world is nothing but folly (I. Cor. i. 20). According to the world, glory consists in heroic courage; but what greater heroes are there than those disciples of the cross who are called apostles and martyrs and saints? Lastly, according to the world, glory consists in being admitted into the intimacy of the great and of monarchs; but the cross admits me into the intimacy of God, of Jesus Christ His Son, of all the angels and all the saints. So is it not incomparably more glorious? Honor, then, to the cross! May it be welcome every time that it presents itself. Honor to crucified souls! They are the favorites of God, His special friends, who wear the liveries of the great King Jesus. Is it thus that we appreciate the cross? Do we not perhaps entertain quite different sentiments, even to the extent of murmuring and complaining when we see it approach?

Resolutions and spiritual nosegay as above.

Thursday in Passion Week.

Summary of the Morrow's Meditation.

We will to-morrow consider the cross as a sacred chair, whence Jesus teaches us: 1st, to know God; 2d, to know ourselves. We will then make the resolution: 1st, to maintain a great respect for God and His infinite perfections, and to testify it to Him by our profound devotion in prayer and at church; 2d, to have a horror for all kinds of sin, and to take to heart the salvation of our soul. Our spiritual nosegay shall be the words of St. Augustine: "*Lord, may I know Thee, that I may love Thee ; may I know myself, that I may hate myself.*"

Meditation for the Morning.

Let us honor the cross of Jesus Christ as being the book of the elect, the science of the saints; it is therein we learn better than by all the books which have ever been written, better than in the schools of all masters whatsoever, what God is and what we ourselves are. Let us thank Our Lord for these lessons.

FIRST POINT.

The Cross Teaches us to Know God.

To know God is not only the first and the most excellent of all kinds of knowledge; it is also the most necessary, for we cannot adore God, respect Him, and abase ourselves before Him, excepting in proportion to the knowledge we have of His greatness; we cannot praise and bless Him, except in proportion to our knowledge of His infinite wisdom; we cannot serve Him by a holy life, except in proportion to the knowledge we have of His infinite holiness; lastly, we love Him only in proportion as we know that He is good. Now, this greatness, this wisdom, this holiness, this goodness, it is the cross which gives us the knowledge and the highest opinions of them. 1st. It teaches us how great God is. Certainly the heavens recount His glory, and the innumerable worlds in the midst of which the whole earth, of which we form so small a part, is less than a drop of water in the ocean, show forth His greatness. Doubtless the prophet Baruch astonishes us when he shows us, at the voice of God, the sun and the moon hastening to place themselves in the spot marked out for them, the stars coming in their turn to say to God, Behold us! and advancing under His orders, like an army ranged in imposing order of battle. Isaias is not less admirable when he makes us see all the nations as so little a thing in the sight of God that they are less than a drop of water glistening upon a rose; they are as though they were not, but the cross always gives me the most lofty ideas of God. There I see a God, the Victim, offered to God by a God-Priest, and I say to myself: If we may judge of the greatness of kings by the excellence of the gifts made to them and by the dignity of those who serve them, O eternal God, how great Thou art, Thou in presence of whom a God so profoundly abases Himself, Thou who hast as Thy minister a God-Priest, and who dost receive from His hands a God-Victim. Yes, Thou art truly infinite in greatness, and Thou canst not conceive an expression greater than what Thou art. 2d. The cross speaks to us of the infinite wisdom of God, and what but infinite wisdom could have inclosed the immense within a limited being, conciliated all kinds of suffering with the beatific vision, make the immortal die, offer to divine justice a satisfaction superior to the offence, and wherein are displayed at one and the same time all the magnificence of mercy? O divine wisdom, who workest such wonders in the cross, thou art, of a truth, infinite! 3d. Yet the holiness of God does not shine in the cross with a lesser splendor. Let us behold it, pursuing in a well-beloved Son sin, even to its merest shadow, punishing the appearances only of it with inflexible severity, and washing them in the very blood of this cherished Son. 4th. What shall we say of the divine goodness, of the goodness of God the Father, who immolates His Son for a rebellious, wicked, ungrateful slave; of the goodness of God the Son, who, entering into the views of His Father, gives Himself up to torments and death in order to save us? Is not this the most sublime ideal of goodness? O divine perfection! O greatness! O wisdom! O holiness! O goodness! how magnificently the cross shows you forth! I have not known you enough up till now; but now I see you are so beautiful, so ravishing that I will consecrate the remainder of my life to adore, to praise, to bless, and to love you.

SECOND POINT.

The Cross Teaches us to Know Ourselves.

I question the cross respecting my nature, and it replies to me that I am a mysterious mixture of greatness and of baseness. What grandeur there is in me! The dignity of my nature is so eminent, that God has redeemed me in preference to the angels, whom He has left without redemption. My salvation is so dear to God that, in order to work it, a God came down from heaven and died upon the cross. My soul is placed in such a lofty position in the esteem of God that, in order to redeem it, He gave the blood of His Son. Sublime truths which teach us to value our salvation above everything else, not to allow our soul, which is so great, to debase itself to earthly and sensual affections, but always to maintain it at the height of its excellence by a pure and holy life. By the side of this greatness the cross shows us our baseness and our misery; it tells us that sin has cast us into so profound a state of wretchedness that it is impossible for us to raise ourselves out of it by our own efforts, incapable even to offer to the offended God the least reparation; it tells us that original sin has deposited in us a tendency to evil, an aversion from what is commanded, a heart so bad, so hard, that a God was unable to gain us except at the price of His death upon the cross, and even then that His success has been very small. Oh, how worth nothing we are! how miserable we are! how humble and penitent we ought to be, how contrite, how mortified! Such are the lessons given us by the cross.

Resolutions and spiritual nosegay as above.

Friday in Passion Week.

Summary of the Morrow's Meditation.

We will meditate to-morrow : 1st, on the sufferings which Mary experienced at the foot of the cross ; 2d, on the virtues which she practised there ; 3d, on the words addressed to her by Jesus. We will then make the resolution : 1st, often to honor the compassion of the Blessed Virgin by pious aspirations ; 2d, to imitate to-day, by some special act, the patience, the humility, and the spirit of sacrifice of which she offers us an example in this mystery ; 3d, to offer great thanks to Our Lord for having given us Mary to be our mother. Our spiritual nosegay shall be the words of the Church : "*O Mary, abyss of love, make me feel thy sorrows ; make me to weep with thee.*"

Meditation for the Morning.

Let us transport ourselves in spirit to the holy mountain of Calvary, to the foot of the cross, beside Mary. Let us salute this mother of dolors as the queen of martyrs, for she will not allow herself to be called by any other name in this mystery. "*Call me not Noemi (that is, beautiful), but call me Mara (that is, bitter), for the Almighty hath quite filled me with bitterness*" (Ruth i. 20).

FIRST POINT.

The Sufferings which Mary Endured at the Foot of the Cross.

All the most cruel torments which the martyrs endured are as nothing in comparison with the anguish which Mary suffered. The martyrs suffered at least only in their bodies, and besides the unction of grace softened and charmed their torments to such a degree that they were seen to triumph joyfully amidst the most cruel tortures; but in Mary it was her very soul which was transpierced with the sword of grief, without the alleviation of any consolation (Luke ii. 35). And what suffering, O my God! If a mother who sees her son expiring before her eyes suffers indescribable agony, what must Mary have felt, she who felt for Jesus a love which nature and grace elevated to the highest degree; nature in showing her in Jesus the most amiable of sons, the most holy, the most perfect, the most accomplished of men; grace in revealing to her in Him a God infinitely good and infinitely amiable (Hymn, *Stabat Mater*); and this beloved Son she was forced to see dragged through the streets of Jerusalem, to the priests, to Pilate, to Herod, everywhere insulted, scoffed at, despised; she was forced to behold Him scourged, crowned with thorns, proclaimed by the people to be worthy of death, and worse than the thief and assassin Barabbas; she was forced to accompany Him to Calvary, ascending the mountain beneath the weight of the cross, exhausted from loss of strength and of blood, covered with wounds and spittle; and she could not give Him any aid! For a mother like Mary, what a martyrdom! She was forced to see Him stretched upon the cross, to hear the blows of the hammers driving the nails into His feet and His hands, to contemplate Him, with all His wounds, lifted up between heaven and earth, agonizing during three hours; she was forced to hear His last farewell, receive His last sigh, without being able to die with Him (Ibid.). And, what was worse still, she suffered the sorrows which she herself caused her Son by her extreme affliction; and all those other quite inexpressible tortures which the heart of her Son suffered at the sight of all the sins committed by the men who are determined to damn themselves spite of all the means of salvation offered to them. O daughter of Jerusalem! to what can we compare the extremity of thy affliction? It is as great as the sea (Lam. ii. 14). Obtain for me grace to compassionate thy sorrows, O my mother! (Hymn, *Stabat Mater.*) It is my duty: 1st, because a son ought to share in the sufferings of his mother (Ibid.); 2d, because he could not love Jesus who could be insensible to His sufferings (Ibid.); 3d, because my sins are at once both the cause and the object of the sufferings of thy Son, and of thine, O afflicted mother! (Ibid.)

SECOND POINT.

The Virtues Practised by Mary at the Foot of the Cross.

1st. She practises there an unalterable patience. She stands up in the midst of the horrible tempest like a rock surrounded by waves, which beat against it without causing it to fall. Neither the abyss into which she is plunged by her grief, nor the spectacle of death, nor the fury of man, nor the rage of demons is able to cast her down. Her demeanor is full of resolution and courage. Without allowing a complaint to escape her, she adores the designs of God in silence and submits to them. Let us look at ourselves in this beautiful mirror of patience, and let us be confounded. It requires so little to cast us down, to make us lose heart, to excite complaints and murmurings! 2d. The humility of Mary is here equal to her patience. A mother whose son is suffering capital punishment is ashamed to show herself; she is afraid lest the ignominy of her son should rebound upon her, and she hides herself; but Mary shows herself and shows herself even at the foot of the cross (John xix. 25). It is there she awaits all the contempt, all the insults that can be heaped upon her, and is happy to be able to taste with Jesus the chalice of humiliation and to drink it down to the dregs. What a lesson for us! 3d. Mary teaches us the spirit of sacrifice. Knowing that the design of God is that Jesus should die to save the world, she enters with her whole soul into the divine decrees. Heavenly Father, she says, take Thy sword, strike the Victim, tear my entrails, wrench out my heart by taking from me my beloved Son. I resign myself to it for the sake of Thy glory and the salvation of the world. What a sublime example of the spirit of sacrifice!

THIRD POINT.

The Words Spoken by Jesus to Mary.

Whilst Mary was suffering such great sorrows, and practising such lofty virtues, Jesus, turning His eyes towards St. John, and seeing in him, state the Fathers, the representative of all the faithful, Woman, He says to Mary, behold thy son; I substitute him to fill My place (John xix. 26). Blessed words, by which Jesus gives us His mother to be our mother, He who had already given us His Father to be our Father, that we might be His brethren, having the same Father and mother! Words which ought to fill our hearts with confidence, with consolation, and with happiness! O Mary, thou art my mother! I no longer fear, I am happy and I hope!

Resolutions and spiritual nosegay as above.

Saturday in Passion Week.

Summary of the Morrow's Meditation.

We will to-morrow resume our meditations upon the cross, considered as the great book which instructs us, and we shall see that it teaches us: 1st, to feel a tender interest in all that has regard to our neighbor; 2d, to despoil ourselves entirely of the spirit of selfishness. Our resolution shall be: 1st, to seek in all things the glory of God and the good of our neighbor; 2d, to detach our hearts from everything else. Our spiritual nosegay shall be the words of St. Paul: "*I judge not myself to know anything among you but Jesus Christ and Him crucified*" (I. Cor. ii. 2).

Meditation for the Morning.

Let us adore Jesus crucified as our Doctor and our Master. It is He who teaches us thoroughly what we ought to seek for esteem, and love; that is to say, the interests of God and of our neighbor, what we ought to fly, despise, and hate; that is to say, everything that is opposed to these two interests. Let us thank Him for this lesson, and let us ask of Him grace to conform our conduct to it.

FIRST POINT.

The Cross Teaches us to Feel a Tender Interest in all that Regards our Neighbor.

The cross, in fact, shows us: 1st, in our neighbor, whoever he may be, a man so tenderly loved by Jesus Christ that, in order to save him, He came down from heaven to earth, became man and gave His blood, His honor, His liberty, and His life, and identified Himself so entirely with each child of Adam as to say, All that is done to the least of My brethren I look upon as done to Myself, and all which is refused to them I look upon as refused to Myself (Matt. xxv. 40-45). Now, this being understood, it is evident that under penalty of failing in our duty to Jesus Christ we ought to feel a tender interest in all that has regard to our neighbor, to his salvation, to his reputation, or to his honor, to his joys, to his sorrows, to his prosperity or his reverses. To be careless respecting the interests of a person so dear to Our Lord, to wound him, to grieve him, to injure or scandalize him, is to wound Jesus Christ Himself in the very apple of His eye. All the interests of this man ought to be as dear to us as those of Jesus Christ; we ought to esteem ourselves happy and honored by all we can perform for His service and lovingly seize every opportunity of doing so. 2d. The cross teaches us to what extent we ought to carry zeal for the interests of our neighbor; for if Jesus Christ on the eve of His death commanded us to love each other as He Himself has loved us (John xiii. 34), the cross offers itself to us as being the commentary on this precept; it teaches us that we ought to be disposed to make the utmost sacrifice for the good of our neighbor, to suffer everything from others without making any one suffer, to bear privation and discomforts, and, according to circumstances, to immolate ourselves wholly for the happiness of our brethren, since it is thus that the crucified Jesus has loved us. Let us here examine ourselves. How many services which we might have rendered have we refused to our neighbor? How many times have we seen him suffering discomfort and embarrassment, compromising his interests by awkwardness or ignorance? We might have freed him from his painful position by a word of good counsel, by charitable advice, by a good office which would have cost us little; and turning away our heads, we have passed by without showing any interest in his misfortunes. Oh, how far we are from loving our brethren as Jesus Christ has loved us!

SECOND POINT.

The Cross Teaches us to Divest ourselves Entirely of the Spirit of Selfishness.

Until Jesus Christ came, no one knew how to live except for self. To obtain for one's self enjoyments, riches, and glory; to keep at a distance from one's self poverty, suffering, and humiliation, such was the whole care of the human race. Jesus Christ appeared on the cross, showed Himself to the world, and from the summit of this new seat He says to the world: Learn of Me to forget yourself, to divest yourself of that miserable egoism which thinks only of self; which troubles itself but little that others should be unhappy, provided that it can enjoy; which believes that it aggrandizes itself by surrounding itself here below with false goods, often even to the prejudice of others, and that it lowers itself by leading a hidden, unknown life, by depriving itself or by suffering in order to oblige others. Behold Me, I am the well-beloved Son of God, and yet I am poor, suffering, humiliated. If riches and abundance, pleasure and glory had been true goods, would not God, My Father, have given them to Me? If poverty, humiliation, and suffering had been evils, would He have made of them My portion? Learn from My example, and know that all which passes away is nothing (Philipp. iii. 8); "*that all is vanity except to love God and to serve Him*" (I. Imit. i. 3). These sublime truths, issuing from Calvary eighteen centuries ago, have changed the face of the world; inspired thousands of souls with the most noble sentiments and the most generous sacrifices for the welfare of religion and of society ; and such souls as these have been seen, detached from everything except the cross, to sell their goods for the solace of the poor, embrace an austere life, that they might belong more certainly to God, submit to persecution as to a piece of good fortune, and return delighted to have been deemed worthy to suffer for Jesus Christ. Behold how the cross has thus withdrawn egoism from the world, and has substituted for it charity with its heroic devotedness. Whosoever does not understand these things is possessed of only a false virtue, an alloy of a semblance of devotion united with the love of self, with a research of what will administer to its tastes and its comforts, frivolity, the love of the world and of its vanities : a worse state than is that of great vices, because great vices awaken remorse, whilst this false devotion makes the soul slumber in a security which leads it to death. Are we not of the number of those who have not yet understood this great lesson of the cross : death to egoism ?

Resolutions and spiritual nosegay as above.

Palm Sunday.

The Gospel according to St. Matthew, xxi. 1–9.

"And when they drew nigh to Jerusalem, and were come to Bethphage, unto Mount Olivet, then Jesus sent two disciples, saying to them :

Go ye into the village that is over against you, and immediately you shall find an ass tied and a colt with her; loose them and bring them to Me, and if any man shall say anything to you, say ye, that the Lord hath need of them, and forthwith he will let them go. Now all this was done that it might be fulfilled which was spoken by the prophet, saying: Tell ye the daughter of Sion: Behold thy King cometh to thee, meek, and sitting upon an ass, and a colt the foal of her that is used to the yoke. And the disciples going, did as Jesus commanded them. And they brought the ass and the colt, and laid their garments upon them, and made Him sit thereon. And a very great multitude spread their garments in the way: and others cut boughs from the trees, and strewed them in the way: and the multitudes that went before and that followed cried, saying: Hosanna to the Son of David: Blessed is He that cometh in the name of the Lord: Hosanna in the highest."

Summary of the Morrow's Meditation.

We will meditate to-morrow upon the gospel of the day, and we will consider: 1st, why Jesus enters triumphantly into Jerusalem, knowing that He is going there in order to be crucified; 2d, what are the characteristics of His triumph. We will then make the resolution: 1st, to renew our love of the good pleasure of God, even when He crucifies us; 2d, to perform our communions better, by joyfully receiving Jesus within us as a conqueror who comes to take possession of our heart. Our spiritual nosegay shall be the words of the prophet: "*Say to the daughter of Sion, Behold thy King cometh to thee, full of meekness*" (Matt. xxi. 5).

Meditation for the Morning.

Let us transport ourselves in spirit before the Saviour entering Jerusalem in triumph; let us join ourselves to the people who acclaim Him, and let us say with them, "*Hosanna to the Son of David: Blessed is He that cometh in the name of the Lord!*" (Matt. xxi. 9.)

FIRST POINT.
Why Jesus Enters Jerusalem in Triumph.

It is a very strange circumstance that Our Lord, who all His life long has flown from glory and splendor, to bury Himself in obscurity, accepts the honors of a triumph with all its demonstrations of public esteem, and that on the eve of His death, when He perfectly well knows that He is about to be crucified. Whence comes this difference in His conduct? Why accept to-day what He had always hitherto refused? It was: 1st, that He desired to show us how He loved the will of His Father. His whole life employed in praising Him, had doubtless been a splendid homage rendered to His adorable will; but a solemn opportunity presents itself of carrying this perfect love up to the point of the most sublime heroism. His Father asks of Him the sacrifice of His liberty, of His honor, of His life. O my Father, behold Me, He exclaims, I come to fulfil Thy commands. I come not with the patience which resigns itself, but with the joy which triumphs, to show the world how amiable is Thy will, above all when it crucifies, Thy good pleasure, ravishing above all when it immolates. 2d. Jesus triumphs because He is about to give us the two greatest proofs of His love; the one at the Supper, in establishing the sacrament of His love; the other on Calvary, in dying for us. For a long period He had desired the one and the other with incredible ardor (Luke xxii. 15; Ibid. xii. 50). The moment so long desired has come, so much happiness is well worth a triumphal progress. Going to the Supper, it is a good Father who comes, His heart overwhelmed with joy, to leave to His children the most magnificent inheritance; going to Calvary, it is a Saviour King who is going to wage war against the infernal powers, the world, the flesh, and sin. It will cost Him all the blood which flows in His veins, even His very life, but it does not signify. He will save us at that price. He is glad, that is why He triumphs. Oh, who is there that will not bless this divine Conqueror, and that will not cry with all the people: " *Hosanna to the Son of David!*" 3d. Jesus triumphs in order to teach us the value of the cross and of sufferings. The world makes happiness to consist in enjoyments which pass away, in honors which fade. In order to disabuse it Jesus took flight when the people wanted to make Him king (John vi. 15). He withdrew into a place apart when He willed to transfigure Himself, and when He was offered enjoyments He stole away from them; but when there was a question of being humiliated and suffering, Arise, let us go! He exclaims (Matt. xxvi. 46), the cross awaits Me; it is My glory. I will go in triumph to seek it, I will bear it on My shoulders as the prophet hath said (Is. ix. 6). A beautiful example which has made twelve millions of martyrs hasten to death, singing canticles of joy! How, after that, can we place our glory in reputation, our felicity in pleasure, our shame in humiliation, instead of saying with the Apostle: "*I please myself in my infirmities, in reproaches, in necessities, in persecutions, in distresses for Christ*" (II. Cor. xii. 10).

SECOND POINT.
Characteristics of the Triumph of Jesus Christ.

1st. It was a triumph which was humble and

full of meekness; Daughters of Sion, says the prophet, thy King comes to thee in a poor and humble condition (Zach. ix. 9), but with ravishing goodness and inexpressible sweetness (Matt. xxi. 5). He is so humble that He has chosen the poor and children to sing His praises; He is so meek that He treats the proud Pharisees with the utmost gentleness when they ask Him to make the multitude cease their acclamations. It was by His poor and simple humility, by His meekness always full of complaisance, that the King of kings is recognized, and it is by these characteristics that His disciples also ought to be known.

2d. The triumph of the Saviour is emblematic of the dispositions with which we ought to receive Him when He comes by Holy Communion, triumphing with love, into our hearts. The raiments laid on the ground beneath His feet are a type of the laying down of the bad habits with which our soul is, as it were, covered. The branches of trees strewed on the ground symbolize the retrenchment of the thousand desires, the attachments, and the self-will, of which Our Saviour asks the sacrifice. The palms which the people carry represent the victories which we ought to obtain over our passions, and which we ought to offer to the Saviour at every communion. Lastly, the cries of triumph with which the air resounds are the symbol of the holy transports with which we ought to receive Him at His arrival in our hearts. Are these the dispositions we bring to our communions?

Resolutions and spiritual nosegay as above.

Monday in Holy Week.

Summary of the Morrow's Meditation.

We will meditate to-morrow on what Christ suffered from His apostles during His Passion; that is to say, 1st, from Judas who betrayed Him; 2d, from St. Peter who denied Him; 3d, from the other apostles who forsook Him. We will then make the resolution: 1st, to mistrust ourselves and to confide in God only; 2d, patiently to bear all the trials which may be inflicted on us by creatures, even by our best friends. Our spiritual nosegay shall be the complaint of Job applied to Our Saviour: "*My kinsmen have forsaken me*" (Job xix. 14).

Meditation for the Morning.

Let us adore Jesus Christ whose heart was so full of love for His apostles, so patient to their faults, so generous in the favors He heaped upon them, and yet, notwithstanding so much kindness, betrayed, denied, and forsaken by them. Let us adore His mercy, let us praise and bless His indulgence towards human weakness.

FIRST POINT.

Jesus Betrayed by Judas.

Our Lord had overwhelmed Judas with His kindnesses; He had made him His apostle and His friend, He had honored him with the power of working miracles at the Last Supper, He had washed his feet, He had given Himself entirely to him in the Holy Communion, and behold, instead of being grateful to Him for so many benefits, the miserable wretch sells Him to the Jews for thirty pieces of silver, walks at the head of His enemies who come to take Him, and gives Him the perfidious kiss which was the signal agreed upon to point Him out to the soldiers who were to arrest Him. Oh, how sad was this treachery to the heart of Jesus! If it be painful, when we love, not to be able to make ourselves loved, what is it, then, to receive in return for our love nothing but perfidy and malice? Our Lord willed to suffer this trial in order to console those who are tried by ingratitude or treachery, and to teach them how to behave in the like circumstances. He meets treachery with nothing but kindness and meekness. My friend, He said to Judas. It was as much as to say to him, if you do not love me any longer, I still love you, and I am as ready to give you pardon as to receive the injury you are doing Me without a cause;

and it is as much as to say to us, never to be angry, even with those of whom we have most cause to complain; to have compassion, rather than indignation, for every man who sins, and never to lose confidence in the divine mercy, since Jesus Christ calls Judas His friend even after his crime. "*Wherefore art thou come?*" (Matt. xxvi. 50) the Saviour adds. How much is contained in this! *Wherefore!* wherefore, O Judas? For thirty pieces of silver and the malediction of God, for a little temporal gain and eternal damnation! What folly! Wherefore, O Christian soul, such anxieties, such earnest solicitude to satisfy pride, ambition, cupidity? What will you gain from it all? Wherefore such pusillanimity in the service of God, tepidity in prayer, time lost in useless conversations, in reading frivolous books? What will you gain from it all? Wherefore your whole life? Wherefore each one of your actions? What is the object of them? What is the fruit of them? Oh, what unreasonableness there is in the man who sins, in the man who proposes to himself any other end except God, whether it be as regards what he does or what he proposes to himself.

SECOND POINT.
Jesus Denied by St. Peter.

Let us leave to the silence of meditation to

show us what, on this occasion, was the sorrow of the heart of Jesus; and let us meditate on that most useful lesson taught us by the fall and the conversion of the apostle.

1st. His fall instructs us. It teaches us, 1st, to mistrust ourselves. St. Peter fell because he presumed on his strength; and thus all who are presumptuous fall when they count upon their own virtue. It teaches us, 2d, not to separate ourselves from Jesus Christ by mingling too much with the world or by dissipation of thought. St. Peter followed the Lord only from afar, says the Gospel. It teaches us, 3d, to avoid all occasions of sin; St. Peter stopped to talk with the servants. It teaches us, 4th, to fortify ourselves by watchfulness and prayer: Jesus Christ had recommended these two means; St. Peter had slept in the Garden of Olives. It teaches us, 5th, to rise promptly after the first fall; because unless we do so we fall from abyss to abyss. St. Peter said at the first assault, "*I know not this man;*" at the second, he confirmed this wretched falsehood by an oath; at the third he confirmed his oath by imprecations (Mark xiv. 71). Thus we fall from one depth into another when we do not hasten to rise.

2d. The conversion of St. Peter instructs us no less than does his fall. It teaches us, 1st, how good Our Lord is; with a single look He pierces the heart of His apostle and converts him. O loving glance! Peter does not seek Jesus; it is Jesus who makes the first advances towards him. Powerful glance! It raises the courage of Peter and makes him shed a torrent of tears. Glance full of sweetness! It spares Peter the shame of his crime, and cures the ulcer without touching it. Generous glance! Jesus forgets His own sufferings in order to occupy Himself with the conversion of His apostle; He comes back to His slave after having been outraged by him. Happy those who, understanding the power of this divine glance, know how to show Him their wounds and open to Him their heart! The conversion of St. Peter teaches us, 2d, to weep over our sins, not from fear, but from love, to weep over them bitterly (Matt. xxvi. 75), to weep over them always. Peter wept until his death over the misfortune of having denied his Master; and his cheeks bore, as long as he lived, traces of the river of tears which flowed from his eyes. Let us collect together in the bottom of our hearts all the lessons offered us by the sin and the conversion of the apostle and let us profit by them.

THIRD POINT.

Jesus Abandoned by the Whole of His Apostles.

The apostles, who had so ardently protested

that they would die for Jesus Christ, lost courage in presence of the danger, and they all abandoned Him. Let us learn from hence: 1st, how weak and miserable man is of himself, and how little is required to make us fail in our best resolutions; how much, consequently, we ought to mistrust our own strength, nor count upon ourselves, nor expose ourselves to occasions of sin, but watch and pray without ceasing, in order to call to our aid His grace, which alone can enable us to live well. Let us learn, 2d, not to count upon the friendships of the world, or to allow ourselves to be disconcerted when they fail us. The apostles had all of them promised Jesus Christ that they would never abandon Him, and at the first signal of danger they all took flight. If Jesus Christ bore this abandonment, let us, following His example, bear to be forsaken by even those on whom we imagined we had the most right to depend; let us be content with having God, who will never abandon us; He will remain with us, and He suffices us.

Resolutions and spiritual nosegay as above.

Tuesday in Holy Week.

Summary of the Morrow's Meditation.

We will meditate to-morrow on what Jesus suffered from His enemies in His Passion, and we shall see: 1st, what were His sufferings; 2d, what were His opprobriums. We will then make a resolution: 1st, heartily to embrace all opportunities of humiliating and mortifying ourselves; 2d, to renounce all pretensions to pride and self-love, as well as all kinds of sensuality. Our spiritual nosegay shall be the words of the Apostle: "*Christ therefore having suffered in the flesh, be you also armed with the same thought*" (I. Pet. iv. 1).

Meditation for the Morning.

Let us adore Jesus Christ teaching us by His example, before leaving the world, to tear out of our hearts the two passions which damn the majority of men: the passion for pleasure and the passion of pride. He combats the passion for pleasure with the most poignant of sufferings, the passion of pride He combats with the most ignominious of humiliations. Let us ask of our divine Saviour pardon for our corruption, the expiation of which has cost Him so dear, and let us thank Him to have been so willing, in order to save us, to submit to so many torments and so much ignominy.

FIRST POINT.

The Tortures which the Enemies of Jesus Christ made Him Suffer.

These men, who carried their inhuman and cruel proceedings to the point of ferocity, did not leave any part of His body untouched by suffering. On the night which preceded His death, they wounded His adorable Face with blows; on the very day of His death they lacerated His flesh with scourges; the blood flowed, His whole body was nothing but one great wound, all His bones were exposed, and His head was crowned with thorns. After having suffered all these tortures, they made Him carry His cross to Calvary, they pierced His hands and feet with nails, they gave Him gall and vinegar to drink. Let us meditate upon these frightful sufferings; let us enter into the thought which inspired the God who suffered them, and who willed thereby to inspire us with the hatred of our own flesh. Who, after meditating on all this, would dare to flatter his body, to humor it, to spare it, to procure pleasure and enjoyment for it? Who would not be determined to mortify it and make it suffer? For we are not Christians excepting under these conditions.

What an examination ought we to make here of ourselves! what a reformation there ought to be effected in our sentiments and our conduct! We love pleasure so much, we are so afraid of discomfort and suffering! How dare we call ourselves Christians?

SECOND POINT.

The Opprobriums which the Enemies of Jesus Christ made Him Suffer.

In the Garden of Olives, Jesus was bound and led from there, like a criminal, to Caiphas in the midst of a thousand insulting cries. On the night which followed His arrest, He was given up to the mercy of His enemies, who wounded Him with blows and with cuffs, who spat in His face, and, after having bound His eyes, showered great blows upon Him, while saying to Him: *Guess who it is that has struck Thee.* On the day which followed this terrible night they march Him through the streets of Jerusalem, covered with the robes of a mock king; they rail at Him and insult Him with being a fool. Brought back from thence to the tribunal of Pilate, He is put in comparison with Barabbas; the whole of the people, who but a little while before had received Him in triumph, proclaim that Barabbas, a thief and assassin, is less guilty than He; and with cries of rage and fury, they demand the death of Him who had never done anything but what was good. Then they crown Him with thorns, they put upon Him a scarlet garment, in imitation of a royal mantle, and they place in His hand a reed by way of sceptre; and all the people rail at Him as being a mock king. Farewell to the renown of His wisdom: He is considered as being only a fool; farewell to the renown of His power: nothing but weakness is visible; farewell to His reputation for innocence and holiness: henceforth, in the opinion of the public, He is nothing more than a criminal, a blasphemer, a man more worthy of death than are thieves and assassins. He is crucified between two thieves, as being the worst amongst them; and all the people gathered together round His cross overwhelm Him, down to His last sigh, with insults and expressions of contempt. Behold how Jesus Christ teaches us humility, submission, dependence; behold how He condemns pride which cannot bear the least contempt, and becomes impatient over the slightest things, and complains at the slightest contradictions; self-love which revolts at seeing the preference given to others, susceptibilities and pretensions; behold how He teaches us to be content with the esteem of God alone, and to count human judgments as nothing, together with public opinion and the vain speeches of those who mock at piety. What fruits have we derived up till now from these divine lessons? What progress have we made in the bearing of a want of consideration for us, words which wound us, things which hurt our self-love? O Jesus, so humble, have pity on us, convert us!

Resolutions and spiritual nosegay as above.

Wednesday in Holy Week.

Summary of the Morrow's Meditation.

We will to-morrow accompany Jesus Christ: 1st, when He ascends the Mount of Calvary; 2d, when He is crucified there. Our meditation on these two mysteries will make us take the resolution: 1st, cheerfully to bear all the crosses of life; 2d, to renew within us the love of Jesus crucified. Our spiritual nosegay shall be the words of a saint: "*My Love is crucified.*"

Meditation for the Morning.

Let us adore Jesus Christ condemned to death at the tribunal of Pilate; let us, in this mystery, admire a mystery of love. Men thought that they were only carrying out their own hatred; they were carrying out the designs of God; they were seconding the love of the Father, giving up to death for us His well-beloved Son; they were seconding the love of the Son, who was glad to die, 1st, in order to save us; 2d, to teach us, **by His example,** to maintain meekness and equanimity amidst the unjust judgments of the world or the trials sent us by Providence. Thanks, O Jesus, for this great lesson! The Jews cried out that Thou didst merit death; that Thou shouldst live no longer (Matt. xxvi. 66; John xix. 15). It is to me, O my Saviour, it is to my vanity, to my sensuality, that these words are applicable. Yes, these passions deserve death; they must live no longer. O Jesus, make them die in me, so that I may love Thee and henceforth live only for Thee!

FIRST POINT.

Jesus Ascending Calvary.

Hardly had the sentence of death been pronounced before the cross was presented to the Saviour, and He was ordered to take it upon His shoulders and carry it to Calvary. Who can express the love with which He seized it: that cross for which He had sighed for so long a time; that cross which was about to save the world and to reconcile earth with heaven; that cross which was about to teach the whole human race patience under trials and the road to Paradise! O cross forever lovely! I see my Saviour bow down His shoulders under thy weight, and set off for the place of execution; I rise and follow after Him, and I say to myself: Could I, after that, drag along my cross impatiently and ill-humoredly? Could I do otherwise than bear it cheerfully, without murmuring and complaining? O cross! whatever you may be, sufferings of the body or sufferings of the soul, come, come to me; I accept you cheerfully; I will bear you henceforth with courage and resolution; I will even add voluntary mortifications, in order more perfectly to resemble my Jesus bearing His cross. It was in meditating upon this mystery that the saints fell in love with the cross; a St. Paul to the extent of calling it a precious grace (Philipp. i. 29); a St. Peter to the extent of saying, Rejoice when you bear the cross with Jesus Christ (I. Pet. iv. 13); a St. Andrew exclaiming, when he beheld the cross on which he was to die, "O good cross, so greatly desired" (*Vita S. And.*); a St. Teresa, who cried out, "*Either suffer or die*"; a St. Catharine of Siena, entreating, "*Not to die yet, but to suffer longer.*" Jesus, during His progress to Calvary, meets: 1st, Mary, to teach us to have recourse to her in all our troubles; 2d, Simon the Cyrenian, to teach us to remember that every Christian may alleviate the weight of the cross of Jesus, whether by diminishing the faults which weigh so painfully upon His heart, whether by bearing in a Christian manner all the crosses which make but one with His; 3d, the daughters of Jerusalem, who weep at seeing the sad state to which He is reduced. "*Weep not for Me,*" He says to them, "*but weep for yourselves*" (Luke xxiii. 28). It is thus, O Saviour, that Thou dost forget Thyself to think only of us; whilst we, alas! so little know how to pity either Thy sufferings or the sufferings of our neighbor. We think only of ourselves, and forget all the rest. May we profit by the lesson Thou didst give us here.

SECOND POINT.
Jesus Crucified.

Arrived at the summit of Calvary, our adorable Saviour is divested of His tunic. This tunic adhered closely to His bleeding body, and in tearing it violently away from Him all His wounds are opened afresh. O mystery of suffering! Behold Him naked like a worm in the face of the whole of the people, who scoff at Him. O mystery of ignominy! He is told to lie down upon the cross, and He places Himself upon that hard bed, whilst blessing His Father that the hour of the sacrifice has come. He is told to stretch out His hands, and then His feet, and He allows them to be transpierced with nails, to expiate the abuse we have made of our hands and our feet, of our affections and our deeds. O mystery of obedience! Then He is raised upon the cross, which is fixed in the ground; the shock renews all His sufferings, the weight of His body enlarges the wounds in the feet and hands; during three hours He remains suspended between heaven and earth. It is the Eternal Father who offers this sacrifice for our salvation; it is the Supreme Master who, from the height of this new chair, teaches the world detachment, poverty, humility, obedience, patience, and resignation or conformity to the will of God. O mystery of love! It is love which immolates itself, which demands in return all the love of our hearts (Prov. xxiii. 26). O Jesus, behold the poor heart which Thou dost ask for; I give it to Thee; fasten it to Thy cross, so that I may say with the Apostle: *"With Jesus Christ I am nailed to the cross"* (Gal. ii. 19). Thou hast said, *"If I be lifted from the earth, I will draw all things to Myself"* (John xii. 32); accomplish Thy word, O Lord, draw me to Thee; draw there all my heart (Cant. i. 3); may it henceforth live only for Thee, may it be Thine alone, in life and at death.

Resolutions and spiritual nosegay as above.

Holy Thursday.

Summary of the Morrow's Meditation.

We will meditate to-morrow upon the two great mysteries which this holy day recalls to our memory; that is to say: 1st, the institution of the Eucharist; 2d, the institution of the priesthood.

We will then make the resolution: 1st, to make the best communion of the year to-morrow; 2d, to pass the whole of the day in a great feeling of gratitude towards Jesus Christ for the institution of the Eucharist and the priesthood. Our spiritual nosegay shall be the words of a holy Abbot: "*O God! prodigal of Thyself through love for us*" (Guerric, Abbot, *in Fest. Pent.*).

Meditation for the Morning.

Let us transport ourselves in spirit to the Last Supper, where Jesus Christ, on the eve of His death, assembled His apostles together, like the good father of a family who, being near his end, assembles his children round his death-bed in order to address to them his last farewells, to inform them of his last wishes, and to leave them the legacy which his love has provided for them. It is then, above all, that he testifies to them how much he has loved them (John xiii. 1). Let us assist with resolution and love at this touching spectacle, and let us meditate on the two great mysteries of the day, the institution of the Eucharist and the institution of the priesthood.

FIRST POINT.

The Institution of the Eucharist.

Let us first admire Jesus Christ on His knees before His apostles and washing their feet, in order to show to all coming ages what profound humility and what perfect charity are required by the sacrament which He is about to institute and they to receive. Then He places Himself at the supper-table, takes bread, blesses it, breaks it, and distributes it among His apostles, saying to them, *"Take ye and eat, this is My body."* Then, in the same way, He takes the cup, and gives it to them, saying, *" Drink ye all of this, for this is My blood of the New Testament which shall be shed for many unto remission of sins "* (Matt. xxvi. 26–28). Oh, who can help recognizing in all this the love of Jesus Christ! The divine Saviour on the eve of quitting us cannot bear to be separated from us; I will not leave you orphans (John xiv. 18), He had said; My Father recalls Me, but in going back to Him I shall not be separated from you. My death is fixed in the eternal decrees, but in dying I shall know how to survive Myself in order to remain with you. My wisdom has found the means, My love is about to carry them into execution. In consequence, He changes the bread into His body, the wine into His blood; and in virtue of the inseparable union of the soul with the body and the blood, in virtue of the indissoluble unity of the divine Person with human nature, what was formerly only bread and wine is now the adorable Person of Jesus Christ, whole and entire, His sacred Person, as great, as powerful as it is at the right hand of the Father, governing the whole universe, adored by the angels who tremble in His presence (Preface of the Mass). This miracle is followed by another. What I have just done, Jesus Christ says, you, My apostles, will also do; I give to you the power, and not only to you, but to all your successors down to the end of time, because the Eucharist, being the soul of religion and its essence, ought to last as long as it itself will last. Such is the rich inheritance which Jesus Christ has provided for His children throughout the course of ages; such is the testament which this good father of a family at the moment of His departure has made in favor of His children; His dying hands have written it and signed it with His blood; such is the benediction which this good Jacob gave to His sons assembled around Him before leaving them. O precious inheritance! Dear and amiable testament! rich benediction! My God! my God! how shall I thank Thee enough for so much love?

SECOND POINT.

Institution of the Priesthood.

It seemed, Lord, as though Thou hadst exhausted all the riches of Thy love towards us, and yet behold, a fresh marvel is revealed. It is no longer only the Eucharist which is given us on this holy day; it is the priesthood with all the sacraments, with the holy Church, with an infallible authority to teach us, power to govern, grace to bless, wisdom to direct. For all that is essentially connected with the Eucharist, either as a preparation disposing the soul to receive it, or as a consequence to preserve and develop the fruits of it. Consequently, Jesus Christ, as the Sovereign Pontiff, was able to bestow, and really did bestow all these powers, by the single word: *Do this.* O priesthood, which dost enlighten, purify, and inflame the souls of men, which dost dispense on earth the mysteries of God and the riches of grace; priesthood, which, helpful to the soul which has fallen as well as to the just, makest repentance to be felt and openest to us heaven, which dost gather together sinners and dost give them back their innocence; priesthood, which dost sustain the wavering soul and enablest it to make progress in virtue; which dost protect the world against itself and against corruption; against Heaven and its vengeance; priesthood, ineffable benefit, I bless thee, and I bless God for having given thee to the world. Alas! what would the world be without thee, without thee who art its sun, its light, and its heat, its consolation, its strength, and its support? O holy Thursday! thrice blessed day, which has procured so much happiness for the children of

Adam, never can we celebrate you with enough piety, fervor, and love.

Resolutions and spiritual nosegay as above.

Good Friday.

Summary of the Morrow's Meditation.

We will consecrate our meditation of to-morrow to the consideration of Good Friday: 1st, as a day of love; 2d, as a day of conversion. We will then make the resolution: 1st, to spend this holy day in recollection and in frequent aspirations of love towards Jesus crucified; 2d, to practise, in honor of the cross, some little mortifications, adding to them the sacrifice which costs us most. Our spiritual nosegay shall be the words of the Apostle: *"The charity of Christ presseth us; judging this, that if one died for all then all were dead. And Christ died for all, that they also who live may not now live to themselves, but unto Him who died for them and arose again"* (II. Cor. v. 14, 15).

Meditation for the Morning.

Let us transport ourselves in spirit to Calvary; let us there adore Jesus lifted up on the cross for our salvation; and at the sight of His body, which is but as one great wound, let our hearts overflow with compassion, gratitude, contrition, praise, and love.

FIRST POINT.

Good Friday a Day of Love.

Let us with loving eyes gaze at the divine crucified Saviour; everything from the sole of His feet up to the crown of His head, from the least movement of His heart up to His deepest emotion; everything constrains us to love Him; everything cries out to us: "*My Son, give me Thy heart*" (Prov. xxiii. 26). His outstretched arms tell us that He embraces us all in His love; His head, which could not repose on aught save the thorns with which it is crowned, inclines towards us, to give us the kiss of peace and of reconciliation; His breast, wounded with blows, rises with the beatings of His heart, which is moved with love towards us; His hands, violently torn by the weight of the body; His feet, the wound in which is enlarged by the weight they have to bear; His bruised face; His veins exhausted of their blood; His mouth parched with thirst; lastly, all the wounds with which His body is covered form as it were a concert of voices which cry out to us: "See how I have loved you." And if we could but penetrate into His heart, we should see it wholly occupied with each one of us, as though He had only each one of us to love, begging mercy for our ingratitude, our lukewarmness, and our sins; soliciting for us the help of all the grace which we have received, and which we shall receive; offering His blood for us to His Father, together with His life, all His interior and exterior sufferings; lastly, consuming Himself in the indescribable ardors of love, without anything being able to turn away His thoughts from it. O love, would it be too much to die of love for so much love! O good Jesus, I will say to you with St. Bernard: "Nothing touches me, nothing moves me, nothing constrains me to love Thee so much as does Thy holy Passion. It is there I gain the most from Thee, it is that which unites me the most closely to Thee, and which attaches me to Thee the most strongly." Oh, what good reason had St. Francis de Sales to say that the Mount of Calvary is the mountain of love; it is there that in the wounds of the Lion of the tribe of Juda faithful souls find the honey of love, and that even in heaven, next to divine goodness, Thy Passion is the most powerful of motives, the sweetest, the most violent, to ravish all the blessed with happiness! And I, after that, O crucified Jesus! could I live any other life than one of love for Thee?

SECOND POINT.

Good Friday a Day of Conversion.

In order to prove to Jesus that I really love Him, I must be converted, that is to say, I must die at the foot of the cross to all which belongs to the old man in me; to all my negligences and all my lukewarmness; all my self-love and my pride, all the effeminacy which is so eager in seeking after comfort and enjoyment, so inimical to everything that annoys or displeases; the susceptibility which is hurt by everything; the spirit of backbiting and calumny which finds continually something to speak against; the frivolity and dissipation and want of application which will not permit the soul to give itself up to recollection; the license of the tongue which pours forth all that is in the mind; lastly, all that is incompatible with the love which Jesus crucified asks of His disciples. We must substitute for these evil inclinations the solid virtues taught by the cross; humility, meekness, charity, patience, abnegation. Jesus asks it of us by all His wounds, as though they were so many tongues. Can I refuse Him? Can I still adhere to my attachments when I see Him naked on the cross, and not make my garment of His nakedness, my livery of His opprobriums, my riches of His poverty, my glory of His confusion, my enjoyments of His sufferings?

Resolutions and spiritual nosegay as above.

Holy Saturday.

Summary of the Morrow's Meditation.

We will meditate to-morrow upon the two great mysteries of the day of which we make profession in the Apostles' Creed : 1st, the burial of the adorable body of Our Lord ; 2d, the descent of His holy soul into limbo. We will gather together the lessons which this double mystery teaches us, and we will make the resolution : 1st, to prepare ourselves to-day with special fervor for holy communion to-morrow ; 2d, to imitate the spirit of humility and detachment preached to us by the burial of Our Lord. Our spiritual nosegay shall be the words of the Apostle : *" You are dead, and your life is hid with Christ in God "* (Coloss. iii. 3).

Meditation for the Morning.

Let us unite in the devotion with which Mary and the beloved disciple, Mary Magdalene and the holy women, received in their arms the body of Jesus Christ, when Joseph of Arimathea and Nicodemus had taken it down from the cross. With what loving emotion Mary looked at His bruised body, contemplated His dislocated limbs, and kissed His wounds ! And the beloved disciple, how he cast himself upon the sacred side on which he had reposed the night before, and seeing it open how he longed to enter into it! And Mary Magdalene, how she embraced His sacred feet where she had obtained her pardon, how she washed them with her tears and dried them with her hair! Let us share in the pious sentiments of these holy souls.

FIRST POINT.

The Lessons Taught us by the Burial of Our Lord.

This mystery teaches us, 1st, how we ought to communicate. After the adorable body was taken down from the cross, Nicodemus brought a hundred pounds of a precious perfume composed of myrrh and aloes to embalm it; Joseph of Arimathea offered a white shroud in which to wrap it and a new sepulchre hewn in the rock wherein to bury it, a sepulchre in which no one had yet been laid; then the entrance to the tomb was secured by a stone, which was sealed with the seal of the public authorities, and soldiers were placed in front of it to guard it. It is thus that, when the body of Our Lord comes to us in holy communion, we ought to embalm it with the perfume of holy desires and the aroma of good works; we ought to present to Him a heart shining with the beauty of innocence, symbolized by the shroud without spot; a resolute will to practise what is right, like the stone of the rock; a conscience renewed by penance; and after holy communion we ought to shut the entrance into our heart as with a stone and a seal, by means of holy recollection, and place there modesty, discretion, watchfulness over ourselves, as vigilant guards, to hinder the precious treasure we have received from being taken from us. Is it thus that we act?

This mystery teaches us, 2d, what are the three characteristics of the spiritual death to which all Christians are called, according to the words of the Apostle: "*Do you also reckon that you are dead*" (Rom. vi. 11); "*You are dead and your life is hid with Christ in God*" (Coloss. iii. 3). The first of these characteristics is to love the hidden life, to be, as it were, dead in regard to all that is said or thought of us, without seeking either to see the world or to be seen by it. Jesus Christ in the darkness of the tomb teaches us this great lesson: Let the world forget us, let it tread us under foot, it matters not to us. We ought not to trouble ourselves any more about it than a dead man would trouble about it. The happiness of a Christian soul is to hide its life with Jesus Christ in God. Our evil nature may spend itself in uttering remonstrances, in desiring to be approved, loved, distinguished, we may let it say what it likes; the more its sensitiveness with regard to being esteemed by others is extreme, the more unworthy is it of it, and the more need is there for it to be deprived of it. Let our reputation be destroyed, let us be esteemed as of no account, let us be spared nothing, let us be held in horror, let it be as Thou willest, O Lord! The second characteristic of the spiritual death is, whilst making use of earthly goods through necessity, not to attach any importance to them, not to take any pleasure in the luxuries or comforts of this life, or in the enjoyments of the table, or in the satisfaction of curiosity, which desires to see and know everything, to be, in a word, dead with respect to the pleasures of the senses. To this second characteristic we must add the abandonment of our whole selves to Providence, an abandonment which makes us as though we were a corpse, permit everything and anything to be done to us, without arguing, without willing or desiring anything, indifferent to all kinds of positions and to all kinds of occupations. When shall I have reached this point, O Lord; when shall I cease to love myself, when will all be dead in me, in order that Thou mayest live in me?

SECOND POINT.

The Lessons which the Descent of the Soul of Jesus Christ into Limbo Teaches us.

This mystery teaches us, 1st, the love of Jesus for man. When He issued from His sacred body, His holy soul might have withdrawn into the bosom of God, to repose there after all its sufferings; but His love for man inspired Him with the resolve to descend into limbo to console the patriarchs, and to announce to them that in forty days He would take them with Him into Paradise. It is thus that the love of Jesus knows no repose. After His death, as well as during His life, He does men all the good He can. Thanks, O Jesus, thanks a thousand times for this eagerness to do us good! This mystery teaches us, 2d, the love which ought to attach us to Jesus. At the sight of this holy soul the just who were detained in limbo could not restrain their transports; they burst forth into canticles of praise, of gratitude, and of love, and their heart gives itself wholly to God, their liberator. Behold our models! Why should we feel less gratitude and love, since Jesus died for us as well as for them, since He loves us as He loved them, and promises to us Paradise as He did to them?

Resolutions and spiritual nosegay as above.

Easter Day.

The Gospel according to St. Mark, xvi. 1-7.

"At that time Mary Magdalene, and Mary the mother of James, and Salome bought sweet spices, that, coming, they might anoint Jesus. And very early in the morning, the first day of the week, they come to the sepulchre, the sun being now risen. And they said one to another : Who shall roll us back the stone from the door of the sepulchre ? And looking, they saw the stone rolled back. For it was very great. And entering into the sepulchre, they saw a young man sitting on the right side, clothed with a white robe : and they were astonished. Who saith to them: Be not affrighted ; you seek Jesus of Nazareth, who was crucified : He is risen, He is not here ; behold the place where they laid Him. But go, tell His disciples, and Peter, that He goeth before you into Galilee : there you shall see Him, as He told you."

Summary of the Morrow's Meditation.

We will consecrate our meditation on the great feast of to-morrow to the consideration of the resurrection of Jesus Christ as the triumph, 1st, of our faith ; 2d, of our hope. We will then make the resolution : 1st, to praise, glorify, and bless the risen Christ by frequent aspirations, *alleluia!* 2d, often to give expression to acts of faith in the divinity of Jesus Christ, in His religion, and in His Church, as well as acts of hope of a future life. Our spiritual nosegay shall be the exclamation of the Church on this great feast: "*Praise and love to the risen Jesus Christ.*"

Meditation for the Morning.

Let us celebrate this morning in praising, adoring, and loving the risen Jesus. Let us rejoice and be thrilled with gladness. This is the day which the Lord has made, the day of victory and of triumph. Let us unite with the angels in singing glory to God, alleluia!

FIRST POINT.

The Resurrection of Jesus Christ is the Triumph of our Faith.

Jesus Christ has really risen again. The apostles who attest it and who sealed with blood their testimony could not have been deceived, seeing that they had conversed with Him during forty days; they could not have wished to deceive us, seeing that their dearest interests in this world and in the next were opposed to such an idea (I. Cor. xv. 19), and besides, Jesus Christ, if He had not risen again, could not have been, in their eyes, anything else but an impostor who had cheated them in predicting His resurrection; they would not have been able to deceive us even if they had wished to do so, since the Roman soldiers, who had been appointed to guard the sepulchre, would not have allowed them to carry away the body. It is therefore quite certain, O Lord Jesus, that Thou didst really rise again; it is really quite certain, therefore, that Thou art the great Almighty God, since a dead man cannot rise of himself

(Rom. i. 4), and that God alone, who is the Master of life and death, is capable of such a miracle. O holy feast of Easter! how dear thou art to me; the resurrection of my Saviour is a guarantee to me of His divinity, and is thereby the complete guarantee of all my beliefs (II. Tim. i. 12); for if Jesus Christ be God, His religion is divine; the Gospel, which is His word, is divine; the sacraments which He has established are divine; the Church which He has founded is divine, and in believing it I am certain of not deceiving myself. In following my faith, I am therefore following an infallible guide; and in making the sacrifices it demands from me, I know I do not lose my pains, and that God will recompense me. In vain the infidel attacks my belief, in vain the nations rage, the Jews cry out scandal, and the Gentiles folly; Jesus Christ risen replies to all, and there is not a single objection which does not fall into pieces against the stone of His sepulchre. What a consolation! what a triumph for faith which has no need of anything except this single fact in order to be fully justified! How just it is to reanimate this faith on this beautiful day, to believe the things which religion teaches us even as though we saw them (Heb. xi. 27), and to show ourselves to be men of faith in our conduct, in our language, at prayer, in church, everywhere and always.

SECOND POINT.

The Resurrection of Jesus Christ is the Triumph of our Hope.

Man, who lives only a little while here below in the midst of many miseries, has need of hope; but let him rejoice to-day by singing with the Church: "*Jesus Christ, my hope, is risen.*" The resurrection of the Saviour is the warrant and the assurance to us of a similar resurrection, which will compensate us for all the troubles of this life. "*Christ is risen from the dead, the first fruit of them that sleep*" (I. Cor. xv. 20), says the Apostle. Therefore, after Him, the others who are dead will also rise again from their ashes. We form with Him one perfect whole, a body of which He is the head, says the same apostle; but the members must follow the state in which is their head. What would a body be of which the head would be on one side and the limbs on another? Would it be suitable for the Holy Spirit thus to have designated, under the figure of a head and of members, Jesus Christ and the faithful, if they were to live separate from one another? As, therefore, we form but one body with Jesus Christ, His resurrection implies ours also, even as ours supposes His; the one is essentially connected with the other. "*If Christ be preached,*" says St. Paul, "*that He arose again from the dead, how do some among you say that there is no resur-*

rection of the dead?" (I. Cor. xv. 12)—a consoling dogma, which forms the triumph of our hope amidst the labors and sufferings of this life; for, if we are destined to rise with Jesus Christ, our tears will be therefore changed into joy, our trials into delights, our poverty into abundance, our confusion into glory, our death into eternal life. "*I know,*" says Job, "*that my Redeemer liveth, and in the last day I shall rise out of the earth, and I shall be clothed again with my skin, and in my flesh I shall see my God, whom I myself shall see and my eyes shall behold. This my hope is laid up in my bosom*" (Job xix. 25-27). "*The King of the world,*" says the second of the Machabees, "*will raise us up, who die for His laws, in the resurrection of eternal life.*" "*These I have from heaven,*" said the third, "*but for the laws of God I now despise them, because I hope to receive them again from Him.*" "*It is better,*" said the fourth, "*being put to death by men, to look for hope from God, to be raised up again by Him*" (II. Mach. vii. 9, 11, 14). Lastly, all the martyrs and all the just have died in this hope, awaiting a new earth and new heavens, where the bodies of the saints will be glorious, impassible, immortal, shining like the sun, agile like spirits, where there will be no more sorrows or tears, but where all will be glory and happiness. O magnificent hope! How thankful we shall then be to have suffered in patience, to have mortified and deprived ourselves of the vain enjoyments of this world!

Resolutions and spiritual nosegay as above.

Easter Monday.

The Gospel according to St. Luke, xxiv. 13-35.

"At that time two of the disciples of Jesus went the same day to a town which was sixty furlongs from Jerusalem, named Emmaus. And they talked together of all these things which had happened. And it came to pass that while they talked and reasoned with themselves Jesus Himself also, drawing near, went with them. But their eyes were held, that they should not know Him. And He said to them: What are these discourses that you hold one with another as you walk, and are sad? And the one of them whose name was Cleophas, answering, said to Him: Art Thou only a stranger in Jerusalem, and hast not known the things that have been done there in these days? To whom He said: What things? And they said: Concerning Jesus of Nazareth, who was a prophet, mighty in work and word before God and all the people; and how our chief priests and princes delivered Him to be condemned to death, and crucified Him. But we hoped that it was He that should have redeemed Israel: and now, besides all this, to-day is the third day since these things were done. Yea, and certain women, also of our company, affrighted us, who, before it was light, were at the sepulchre, and, not finding His body, came, saying that they had also seen a vision of angels, who say that He is alive. And some of our people went to the sepulchre, and found it so as the women had said, but Him they found not. Then He said to them: O foolish and slow of heart to believe in all things which the prophets have spoken! Ought not Christ to have suffered these things, and so to enter into His glory? And beginning at Moses and all the prophets, He expounded to them in all the Scriptures the things that were concerning Him. And they drew nigh to the town whither they were going, and He made as though He would go further. But they constrained Him, saying: Stay with us, because it is toward evening, and the day is now far spent. And He went in with them. And it came to pass, whilst He was at table with them, He took bread, and blessed and broke, and gave to them; and their eyes were opened, and they knew Him, and He vanished out of their sight. And they said one to the other: Was not our heart burning within us whilst He spoke in the way, and opened to us the Scriptures? And rising up the same hour, they went back to Jerusalem, and they found the eleven gathered together, and those that were with them, saying:

The Lord is risen indeed, and hath appeared to Simon. And they told what things were done in the way, and how they knew Him in the breaking of bread."

Summary of the Morrow's Meditation.

We will meditate to-morrow on the touching recital of the disciples of Emmaus, contained in the gospel of the day, and we shall see: 1st, what, in this circumstance, were their faults and their virtues; 2d, what the great goodness of Christ was towards them. We will then make the resolution: 1st, to keep ourselves united in Jesus Christ by recollection, and to be docile to the inspirations of His grace; 2d, to watch over our conversations, in order not to allow a word worthy of reproach to escape our lips. Our spiritual nosegay shall be the words of the apostles: "*Was not our heart burning within us whilst Jesus spoke in the way, and opened to us the Scriptures?*" (Luke xxiv. 32.)

Meditation for the Morning.

Let us transport ourselves in spirit to the road leading to Emmaus; let us consider Jesus Christ drawing near to the two disciples, who were travelling, and joining with them in a holy conversation. Let us bless Him for His assiduous charity, and let us beg Him to enable us to profit by this amiable interview.

FIRST POINT.

The Defects and the Virtues of the Disciples of Emmaus.

1st. These disciples did not understand how to wait God's own time. Jesus Christ had said, I will rise again on the third day, and they did not wait until the end of the third day, but set off on their journey filled with discouragement. This is a fault which we also often commit; we want to be heard at the very moment; every delay disconcerts us and shakes our faith. We will deserve that Jesus should say to us as He did to them, "O men of little faith, how slow your heart is to believe!" 2d. They seek their consolation from exterior things, by going a journey to Emmaus. They forget that true consolation is to be found in God alone, and that there is more loss than profit in seeking it from creatures. If Jesus Christ had not hastened to their aid they would have lost their faith, since they had not believed either the holy women or the apostles attesting to them the resurrection of Jesus Christ; they were on the point of losing their hope, seeing that they were already beginning not to hope. "*We hoped*" (Luke xxiv. 21), they said. Lastly, they were about to lose their charity, because they no longer saw in Jesus Christ anything more than a prophet, and no longer spoke of being His disciples, but as being strangers. 3d. It was repugnant to them to understand the connection of two things as inseparable as are the means and the end, that is to say, the cross and glory, death and life, suffering for a short time and enjoying eternally ; and it was necessary that Jesus Christ should recall to their remembrance that important truth. Was it not necessary that the Christ should suffer, and that He should thus enter into His glory ? Are we not a little like them ?

But if these disciples had their defects, they had also virtues which are suited to edify us. Thus, 1st, their conversation is holy ; and to the question put to them by the Saviour : *"What are these discourses that you hold one with another ?"* (Luke xxiv. 17) they were able to reply : We are speaking of Jesus (Ibid. 19). Alas, if the Saviour were to present Himself to us in the midst of our conversations, and were to ask us: What are you speaking of? should we not have to blush over many calumnious words, scoffings, disputes, frivolity, ill-temper? and could not Our Lord say to us : Are these the speeches of a Christian, of a heavenly-minded man who aspires after holiness, of a servant of Jesus Christ who has his tongue still tinged with His blood? are these the speeches which at the hour of death you will be very glad to have uttered ? 2d, our pilgrims listen with great respect to the teaching of Jesus Christ ; they engrave it in their heart, which is inflamed with a holy ardor (Ibid. 31); 3d, they attach themselves to Him, and they desire never to be separated from Him any more. Remain with us, Lord (Ibid. 29), they say to Him. Beautiful words, which we ought to address to ourselves! Remain with us in our troubles, to preserve us from impatience, murmurs, and discouragement, and to teach us to bless God in all things; remain with us in temptations and trials, to sustain us; remain with us in times of dryness and disgust; in seasons of sickness and when in danger of death, to assist us; remain with us in the midst of the ills of the Church and of the darkness of iniquity which covers the earth, to defend and enlighten us; 4th, they recognize Our Lord in the breaking of the bread (Luke xxiv. 35), that is to say, in communion; it is there indeed that the Christian soul recognizes all the love of the divine Saviour; 5th, after having received Him they leave for Jerusalem, in order to announce Him to the apostles (Ibid. 33); when we love we have it at heart to make others love those whom we love.

SECOND POINT.

The Touching Kindness of Jesus towards the Disciples of Emmaus.

Jesus Christ takes pity on these two wandering sheep who had separated themselves from the other apostles; He draws nigh to them, addresses them gently, engages in conversation with them whilst walking beside them at the same pace, neither quicker nor slower; He asks them of what they are speaking, not because He is ignorant of it, but that He may afford them an opportunity of opening their hearts to Him, and He Himself makes use of the opportunity to explain to them the mystery of His suffering and death. He reproves them charitably, in order to make them examine themselves and recognize their faults; He proves to them that what was said in the Holy Scriptures from the time of Moses down to the prophets of the Messias is realized in His own person, and at the same time that He enlightens their intelligence He touches their heart, inflames their will, and lights in it the sacred fire of divine love. Lastly, on their arrival at Emmaus, after having allowed them to imagine He was about to pass on, in order to excite in them the desire to keep Him with them, He halts at their hostelry, and as though it had been a church, He there consecrates the Eucharist, distributes it to them, and does not leave them until after having nourished them thus with the Bread of angels. Could there be greater goodness and sweetness? It is thus that Our Saviour acts with regard to ourselves. His predisposing grace comes to seek us in the path of life; it accommodates itself to our weakness, it enlightens us with its divine light, it attracts us by its divine inspirations, it mingles together encouragement and reproaches; lastly, it does not quit us until it has gained us, taking possession of our will without restraining our liberty. Oh, how so much goodness well deserves all our love! How do we respond to it? Are we not unfaithful to grace, and do we not frequently rebel against its inspirations?

Resolutions and spiritual nosegay as above.

Easter Tuesday.

The Gospel according to St. Luke, xxiv. 36–46.

"Now whilst they were speaking these things, Jesus stood in the midst of them, and saith to them: Peace be to you; it is I, fear not. But they, being troubled and frighted, supposed that they saw a spirit. And He said to them: Why are you troubled, and why do thoughts arise in your hearts? See My hands and feet, that it is I Myself; handle, and see: for a spirit hath not flesh and bones, as you see Me to have. And when He had said this, He showed them His hands and His feet. But while they yet believed not, and wondered for joy, He said: Have you here anything to eat? And they offered Him a piece of a broiled fish and a honey-comb. And when He had eaten before them He opened their understanding that they might understand the Scriptures and that it behoved Christ to suffer, and to rise again from the dead the third day."

Summary of the Morrow's Meditation.

We will meditate to-morrow upon the apparition of Jesus Christ to His apostles assembled together at Jerusalem, and we shall see: 1st, the esteem which the risen Jesus had for His sacred wounds; 2d, the esteem we also ought to have for our own sufferings. We will then make the resolution: 1st, often and lovingly to kiss our crucifix and above all the sacred wounds imprinted on it; 2d, cheerfully to accept all the trials of life. Our spiritual nosegay shall be the words of Our Lord: "*See My hands and feet*" (Luke xxiv. 39).

Meditation for the Morning.

Let us transport ourselves in spirit into the midst of the apostles. Let us lovingly kiss the wounds in His feet, in His hands, in His sacred side; let us beg of Him to let grace flow down upon us from them, and to raise us up also to a new life.

FIRST POINT.

The Esteem in which the Risen Jesus held His Wounds.

It is not only during the week of sufferings and on the cross that Jesus Christ presents His wounds to our meditation; He shows them also to us during the week of paschal joys, but with this difference, that last week these wounds appeared to us bleeding and painful, whilst to-day they appear to us as glorious and shining with the rays of the divinity. Jesus Christ willed to preserve them in His risen body: 1st, as being an irrefragable proof that it was really the same body which had suffered for us; 2d, as being the glorious marks of the victory which He had gained over the enemies of God and of the salvation of men; 3d, as the insignia of His love for us, which He delights to show to heaven and earth in order to inflame our hearts with reciprocal love; 4th, as so many divinely-eloquent mouths which plead our cause before His Father and which ceaselessly address to Him, in our favor, an all-powerful prayer; 5th, as sacred fountains out of which we may draw grace continually, with an unlimited confidence in their merits. O divine wounds, so dear to the heart of Jesus, of which you open the doors, how beautiful you are! It is you which cause God to be eternally blessed by all the angels and all the saints, who delight to sing the evangelical canticle: Behold how God has loved man (John xi. 36); it is you who, on the great day of judgment will confound those who have not willed to profit by the benefit of the Redemption (John xix. 37). O adorable wounds! I revere you and I love you. You command me to behold you: I contemplate you lovingly; you are my refuge: I repose in you; you are my light: I instruct myself in your school; you are my strength: you will sustain me in my discouragements; you are furnaces of love: I will approach you, I will keep near to you by an humble, affectionate, assiduous meditation, and I shall be warmed, for you cannot keep near to a great fire without feeling the heat of it!

SECOND POINT.

The Esteem which we ought to have for our own Sufferings.

Whether we will it or not, we must suffer: suffer in our body, suffer in our mind, suffer in our heart; suffer from others who displease and molest us; suffer from ourselves, from our inexplicable fits of sadness, of impatience, of melancholy, and of bad temper; suffer from all human things: sometimes from the death of persons who are dear to us, sometimes from a reverse of fortune, sometimes from the failure of an undertaking, of a humiliation which we have received or have imagined. Now, these sufferings which are the inevitable lot of our humanity ought to be highly esteemed by us: 1st, because Jesus Christ has said: Blessed are those who suffer, blessed are those who weep; 2d, because our divine Saviour has glorified them in His own person, deifying and rendering adorable His very wounds, which enabled Him to merit the glory of His body, His resurrection, His ascension, His repose at the right hand of His Father, and the honor of judging at the last day the living and the dead ; 3d, because without suffering there is no virtue, no merits, consequently no recompense, no salvation ; we become attached to this world and forget heaven, we think only of enjoying the present moments and we do not occupy ourselves with our eternity ; whilst, on the other hand, suffering, borne in a Christian manner, is the source of merits, conduces to the practice of virtues, is the warrant and the measure of the happiness of heaven, so that we deem to be the most beautiful days of our life those in which we have most suffered (Ps. lxxxix. 15) ; 4th, because suffering endured with patience renders us dear to the heart of God the Father, who then sees in us a resemblance to His divine Son. It brings Him near to us to console or relieve us (Ps. xc. 15), for, says the Psalmist, He stretches His paternal hand over the just, weighed down beneath the cross, in order to sustain him (Ps. xxxvi. 24). Daniel is cast into the lions' den, the children of Babylon into the furnace, Joseph into prison ; God is there that He may save them ; 5th, because suffering has always been the delight of the saints. I take pleasure, said St. Paul, in afflictions, whether they be infirmities which attack my body, or calumnies which attack my honor, or poverty which reduces me to be badly-lodged, ill-clothed, badly fed, or persecutions from without, or troubles from within (II. Cor. xii. 10), for it is then that the virtue of Christ dwells in me (Ibid. 9). Either suffer or die, said St. Teresa; I cannot live without the cross; so entirely has Jesus Christ in taking suffering upon Him robbed it of its bitterness and embalmed it with His divine sweetness. Now, in what degree do we esteem suffering? How do we bear what annoys us? Let us beg of Our Lord to give us more Christian sentiments.

Resolutions and spiritual nosegay as above.

Wednesday in Easter Week.

The Gospel according to St. John, xxi. 1–13.

"After this Jesus showed Himself again to the disciples at the sea of Tiberias, where they had fished all the night and had caught nothing. And when the morning was come, Jesus stood on the shore: yet the disciples knew not that it was Jesus. Jesus therefore said to them: Children, have you any meat? They answered Him: No. He saith to them: Cast the net on the right side of the ship: and you shall find. They cast therefore; and now they were not able to draw it for the multitude of fishes. The disciple then whom

Jesus loved said to Peter: It is the Lord. Simon Peter, when he heard that it was the Lord, girt his coat about him and cast himself into the sea. But the other disciples came in the ship. As soon then as they came to land, they saw hot coals lying, and a fish laid thereon, and bread. Jesus saith to them: Bring hither the fishes which you have now caught. Simon Peter went up and drew the net to land, full of great fishes, one hundred and fifty-three. And although there were so many the net was not broken. Jesus said to them: Come and dine. And none of them durst ask Him: Who art Thou? knowing that it was the Lord. And Jesus cometh and taketh bread, and giveth them, and fish in like manner."

Summary of the Morrow's Meditation.

We will meditate to-morrow upon the apparition of Jesus Christ to His apostles on the borders of the lake of Tiberias, as it is recounted in the gospel of the day, and we shall see: 1st, what Jesus Christ did for His apostles in this apparition; 2d, what the apostles did for Him. We will then make the resolution: 1st, in our relations towards our neighbor to imitate the charity of Jesus Christ in this circumstance; 2d, to bring to the services of God the courage of the apostles and their docility to grace. Our spiritual nosegay shall be the words which St. John applied to

Jesus Christ at that time, "*It is the Lord*" (John xxi. 7).

Meditation for the Morning.

Let us adore Jesus Christ impelled by love to show Himself to His apostles on the shores of the lake of Tiberias. Let us thank Him for this delicate attention bestowed on His dear disciples, and let us ask Him to give us an abundant share in the grace of this mystery.

FIRST POINT.
What Jesus did for His Apostles in this Apparition.

The apostles having fished all night without taking anything, had nothing on which to subsist. Jesus takes pity on their distress and comes to them. "*Children*," He says to them, "*have you any meat?*" (John xxi. 5.) What amiable forethought! What paternal solicitude in this Divine Saviour! He is about to provide for their wants, but let us see on what condition: it is on condition that they will labor. For labor is the law imposed on all the children of Adam, and idleness is their ruin. "*Cast the net,*" He said to them, "*on the right side of the ship*" (John xxi. 6). Mysterious words, which signify that in all our actions there is a good and a bad side; the essential thing is to choose the right side. The good side is the side of God, and not that of the creature.

We ought always to consider God alone, without seeking ourselves or paying any attention to human opinions. The good side is the side of grace, and not that of nature ; we ought not to be led to anything by natural inclinations, but by the movements of grace, which alone ought to regulate our whole conduct, our recreations, and our repose, as well as our affairs and our employments. The good side is the side of heaven and not of earth ; we ought to govern ourselves by eternal maxims, like heavenly-minded men, who do not touch the earth except from pure necessity. The good side, lastly, is the side of the cross, and not that of delights and of pleasure ; we ought to attach ourselves to the cross, which is the portion of the elect, and not to the enjoyments of this life. What blessings we lose from want of observing these holy rules! Whilst the apostles were executing the command which had been given them, Jesus lights the fire, cooks the fish, lays the table, puts bread upon it, and when all is ready, "Come and dine," He says to them. They come, bringing to shore their net which contained one hundred and fifty-three great fishes ; and with His divine hands He Himself serves His dear disciples. Who would not admire the charity of Jesus Christ in this circumstance? a foreseeing charity, which could not bear to see His disciples suffer without relieving them ; a generous charity, which, in order to render them a service, condescends to perform with delight the most humble functions; an amiable charity, which studies how to give pleasure to our neighbor.

SECOND POINT.

What the Apostles did for Jesus Christ in this Apparition.

Four things are worthy of remark in the conduct of the apostles: 1st, they immediately obey the command of the Saviour; they cast their net where Jesus had told them, and the fishes were gathered together there. They, who until then had taken nothing, catch at one single throw, as soon as they obey, one hundred and fifty-three great fishes. Let us imitate them; let us always be docile to grace; let us do all through obedience, from the desire to please God, in the manner in which God wills, and we shall be blessed in all our works. 2d. The apostles did not at first recognize Jesus Christ; it was necessary in order to do so that they should have a special grace and a special light, and how few there are who render themselves worthy of this grace! how few who apply themselves to recognize Jesus Christ in His mysteries, in His doctrine, and in His love; how few see and recognize His hand in all events, be they good or unhappy. To know Jesus is the science of the saints, it is the privilege of love and purity, as we may see by the example of

St. John, who was the first of the apostles to recognize His good Master, and to exclaim: "*It is the Lord.*" 3d. On hearing these words of the virgin disciple, Peter girds his coat about him and casts himself into the water that he may sooner reach the feet of Jesus. The ardor of his desires makes him oblivious of danger and difficulty. Fervent hearts spare themselves nothing. As soon as there is a question of serving God, they devote themselves and throw themselves forward, whilst the cowardly and lukewarm hesitate, are wanting in resolution, and are afraid of trouble. 4th. The apostles, during their repast, behave in a respectful manner, they adore, they admire, they enjoy in silence the sweetness of the conversation and the looks directed upon them by Jesus; but no one dare ask Him: "*Who art Thou? knowing that it was the Lord*" (John xxi. 12). It is thus that faithful souls act; the goodness of Our Lord so confounds and humiliates them that they dare not interrogate Him or put vain questions to Him or indulge in vain researches, knowing that it is the Lord, all of whose conduct requires from us veneration and love. The nearer they are to Him, the more respect they have for Him; and if sometimes they ask, *Who art Thou?* it is only with the object of knowing Him better, that they may humble themselves for their littleness in presence of such greatness.

O Lord, I dare not lift my eyes to look at Thee, or open my mouth to speak to Thee! I am nothing but a miserable worm crawling to Thy feet in the dust, poorer and more miserable than I can understand; I am nothing, I can do nothing. Thou alone art good, just, and holy; pour down on me Thy infinite mercy.

Resolutions and spiritual nosegay as above.

Thursday in Easter Week.

The Gospel according to St. John, xx. 11-17.

"Mary stood at the sepulchre without, weeping. Now as she was weeping, she stooped down and looked into the sepulchre, and she saw two angels in white, sitting, one at the head, and one at the feet, where the body of Jesus had been laid. They say to her: Woman, why weepest thou? She saith to them: Because they have taken away my Lord, and I know not where they have laid Him. When she had thus said she turned herself back and saw Jesus standing; and she knew not that it was Jesus. Jesus saith to her: Woman, why weepest thou? Whom seekest thou? She, thinking that it was the gardener, saith to him: Sir, if thou hast taken Him hence, tell me where thou hast laid Him, and I will take Him away. Jesus saith to her: Mary. She, turning, saith to Him: Rabboni (which is to say, Master). Jesus saith to her: Do not touch Me, for I am not yet ascended to My Father, but go to My brethren and say to them that I ascend to My Father and to your Father, to My God and your God."

Summary of the Morrow's Meditation.

We will meditate to-morrow upon the apparition of the risen Christ to Mary Magdalene, as it is recounted in the gospel of the day, and we shall see, 1st, the ardent love of this holy soul in seeking the Saviour; 2d, the manner in which Jesus responds to her love. We will then make the resolution: 1st, often to make, during the day, acts of love towards Our Lord; 2d, every time the clock strikes to animate ourselves to live better, and better to perform the present action. Our spiritual nosegay shall be the words of Wisdom: "*Wisdom is found by them that seek her*" (Wis. vi. 13).

Meditation for the Morning.

Let us adore Jesus Christ granting to St. Mary Magdalene the favor of being the first, after the Blessed Virgin, to whom He appeared, after issuing from the tomb. Let us congratulate this illustrious lover of Our Lord and like her thank Jesus Christ by saying, Good Master. Oh, how good He is, and how He does indeed merit our whole heart's love.

Thursday in Easter Week.

FIRST POINT.

The Ardent Love shown by Mary Magdalene in Seeking the Saviour.

After the death of Jesus, Mary Magdalene seemed not to be able to separate herself from Him to whom she had given all her love; she runs to the tomb, and, finding that the sacred body is no longer there, she imagines that it has been taken away. Where has it been put? She is determined to discover it, no matter at what price; and instead of going away, as the disciples and the other women had done, she remains there, retained by love, in order to seek Him whom she has lost; kept there by grief, to weep over Him whom she cannot find. She remains on the spot, without fearing anything, for, after having lost Jesus, there is no longer anything for her to lose. Jesus was the life of her soul, and having lost Him, it was more desirable in her estimation to die than to live, for she hoped that she should find, in dying, Him whom she could not find whilst living. She remains there, and looks into the sepulchre several times to see if Jesus is not in it. Wherefore do you weep? said the angel who was seated there. "*They have taken away my Lord,*" she replies, "*and I know not where they have laid Him*" (John xx. 13). She turns her head and perceives a man; it is Jesus, who presents Himself to her without making Himself known. "*Sir,*" she exclaims, "*if thou hast taken Him hence, tell me where thou hast laid Him, and I will take Him away*" (John xx. 15). An ardent desire will not admit that anything is impossible, and renders a person capable of everything. How admirable is the love of Mary Magdalene! and how ardent it is! How intrepid is the desire which consumes her to find Jesus! Happy the soul who loves Jesus to the extent of thus desiring Him! God makes our desires the measure of His benefits; and often, with Him, the greatest blessings cost nothing more than a desire. If He sometimes defers granting our petitions at the very moment we offer them, it is only to make us the more earnestly desire His graces, and to make us appreciate them better when He does give them to us. Oh, if we did but desire to possess Jesus within us by recollection and love—I do not say as Mary Magdalene desired Him, but only as much as the worldly man desires wealth and honors—how quickly should we become saints! Our great misfortune is not to love, and, consequently, not to desire ardently our perfection. We lose a trifle, and grieve over it; we lose Jesus in losing recollection, humility, patience, mortification, charity, and it does not in the least distress us, and we do not say with Mary Magdalene: Tell me where He is; I am ready to do all and everything to recover Him. Let us beg of Our

Saviour to infuse into our hearts the ardent desires which would make us saints.

SECOND POINT.

How Jesus Responded to the Love of Mary Magdalene.

St. Mary Magdalene, at the beginning, had only a very imperfect faith, because, not having found Jesus Christ, she supposed that He had been taken away, and not that He had risen. Jesus, however, being touched by her love, sends to her, 1st, two angels clothed in white, whom she sees seated in the very place where His body had been! the one at the head, the other at the feet; then He presents Himself to her in person, beneath the humble form of a gardener. She does not recognize Him, but He makes Himself known to her by one single word: *Mary!* He says to her. Then Mary Magdalene cannot contain herself any longer. Intoxicated with joy and with love, she falls at the feet of Jesus, exclaiming: *Rabboni! good Master!* She would love to remain there forever, kissing His sacred feet, pressing them to her lips and her heart. No, said Jesus, you must do something more than delight in My presence; you must go quickly and find My brethren, and tell them that I am risen, and that soon they will see Me ascend to My Father and your Father, to My God and' your God. Happy Mary Magdalene! she is the first, after

Mary, to whom Jesus has shown Himself; she is the elect of the Saviour, that she may be the apostle of the apostles themselves, and go to announce to them that Jesus is risen. She promptly obeys the command, and teaches us by her example that we must know how to leave Christ, that we may console and help our neighbor; that it is better to be obedient and humble than to enjoy divine consolations; that it is not enough to love—that we must make God, whom we love, to be also loved by others; lastly, that we must know how to moderate our joy, however holy and spiritual it may be, and never abandon ourselves wholly to it, lest we should be tempted to commit some want of respect, which would make us forget the reverential fear which is due to God and the prudent apprehension of losing the graces we have received. What precious lessons are conveyed to us in this behavior of Mary Magdalene!

Resolutions and spiritual nosegay as above.

Friday in Easter Week.

The Gospel according to St. Matthew, xxviii. 16–20.

"And the eleven disciples went into Galilee, unto the mountain where Jesus had appointed them. And seeing Him, they adored, but some doubted. And Jesus, coming, spoke to them, saying: All power is given to Me in heaven and in earth. Going, therefore, teach ye all nations, baptizing them in the name of the Father, and of the Son, and of the Holy Ghost; teaching them to observe all things whatsoever I have commanded you: and behold, I am with you all days, even to the consummation of the world."

Summary of the Morrow's Meditation.

We will meditate to-morrow on the apparition of Jesus Christ to His apostles upon a mountain of Galilee, and we will make of the three portions of His discourse on that occasion the three points of our meditation. We will then make the resolution: 1st, often during the day to make acts of faith in the infallibility of the Church, and acts of love towards Jesus Christ, who, from love for us, gave it this blessed privilege; 2d, to keep ourselves recollected and united with Jesus Christ, who desires that we should be always with Him, even as He is always with us. Our spiritual nosegay shall be the words of Our Saviour: "*Behold, I am with you all days, even to the consummation of the world*" (Matt. xxviii. 20).

Meditation for the Morning.

Let us adore Jesus Christ at the approach of His ascension collecting His apostles together upon a mountain of Galilee, and there giving them the mission to preach the gospel to all the nations, and to their successors the mission to preach it to the end of the world. Let us thank Him for this mission, which interests in the highest degree the whole universe and all coming ages. Let us admire the power and goodness which is shown herein. Let us praise the power, let us love the goodness.

FIRST POINT.

All Power, said Jesus Christ, has been given Me in Heaven and upon Earth.

What words, and what other than a Man-God could utter them! We have indeed often seen men invested with great power, but it has always been a limited power. Who has ever yet had power over the heavens, and over the winds, over thunder and storms, over sicknesses and death? Jesus Christ alone has had unlimited power in heaven and over the earth; the power of commanding the elements, of acting according to His own will over all creation. Of Him alone the astonished people have been able to say: "*Who is this whom the winds and the waves obey?*" O almighty power of my Saviour! I adore and I bless thee with my whole heart. I tremble for the sinners who do not tremble in presence of Thy omnipotence, O Jesus; but I rejoice for the just who serve Thee, for the apostles whom Thou dost send to the conquest of the world. They are indeed lambs in the midst of wolves, but what does that signify? They have nothing to fear, seeing that Thy omnipotence accompanies them. Thou wilt defend them against all assaults; Thou wilt sustain them in difficulties and trials. O all-powerful and good Master! what a consolation it is to see Thee at the head of Thy Church! How much of honor and respect do we not owe Thee for what Thou art in Thyself; of gratitude and love for what Thou hast done in the past; of confidence for what Thou wilt do in the future!

SECOND POINT.

Go then, said Jesus Christ, teach all Nations, Baptizing them in the Name of the Father, and of the Son, and of the Holy Ghost, and Teaching them to Observe all that I have Commanded you.

These words are the consequence of those which preceded them; it is as though Jesus Christ had said: In virtue of and with the aid of the omnipotence which has been given Me, go and teach all nations, convert all the peoples, and extend My empire to the very extremities of the earth. The enterprise is, I know, beyond your strength, but not beyond My omnipotence. Doctors and depositaries of My doctrine, by you will be preserved in My Church the faith which makes saints. Blessed are ye to be chosen for so beautiful a mission, one which glorifies God and which saves souls! If, said St. Catharine of Siena, any one could see the beauty of a single soul, he would die a hundred times a day in order to save it. May we esteem, at this lofty price, souls, and the mission of saving them! "*Baptize the nations,*" continued Jesus Christ, "*in the name of the Father, and of the Son, and of the Holy Ghost.*" Thanks, Lord, thanks for this revelation of the Most Holy Trinity—a revelation the most clear and the most precise which until then had ever been made to the world; thanks for this institution of baptism, which makes of each one of us a child of God, an heir of heaven, and one of Thy members; thanks for all the sacraments, of which baptism is, as it were, the door, and which are the marvellous channels through which Thy grace flows upon us. "*Teach all nations,*" again said the Saviour, "*to observe all that I have commanded you.*" I understand it, my God; faith without works is of no avail; faith is the torch which directs (Ps. cxviii. 105); it shows us what we must do; but we cannot be saved except inasmuch as we really do it, and that we act from motives of faith; for works without faith are of no more merit for salvation than is faith without works. Let us learn from hence always to perform our works so as to be in harmony with our faith.

THIRD POINT.

Behold I am with you all Days, even to the Consummation of the World, said Jesus Christ in Terminating His Discourse.

What magnificent words! They are the titles of the Church to infallibility; for if Jesus Christ assists it "all days" in its teaching, it cannot be deceived; he who listens to the Church listens to Jesus Christ. O delicious consolation! a God makes Himself the warrant of my belief. But it is not only whilst they teach that Jesus is with His apostles; He is even now with us all, through the love He bears us, and which follows us everywhere. To have Jesus Christ with us! What good the thought does to the heart which loves! Can one be in more delightful society? What good it does to the heart which feels itself to be weak! What have we to fear when we have His omnipotence with us? To have Jesus Christ with us in the blessed Tabernacle, where we can visit Him every day, speak to Him, expose to Him our joys and our sorrows; to have Jesus dwelling in the bottom of our hearts if we desire to find Him there, to enjoy Him by means of the practice of the interior life, of recollection, and of love: what matter of confidence and peace!

Resolutions and spiritual nosegay as above.

Saturday in Easter Week.

The Gospel according to St. John, xx. 1–9.

" And on the first day of the week Mary Magdalene cometh early, when it was yet dark, unto the sepulchre; and she saw the stone taken away from the sepulchre. She ran, therefore, and cometh to Simon Peter, and to the other disciple whom Jesus loved, and saith to them : They have taken away the Lord out of the sepulchre, and we know not where they have laid Him. And they both ran together, and that other disciple did outrun Peter, and came first to the sepulchre. And when he stooped down he saw the linen cloths lying: but yet he went not in. Then cometh Simon Peter, following him, and went into the sepulchre, and saw the linen cloths lying, and the napkin that had been about His head not lying with the linen cloths, but apart, wrapt up into one place. Then that other disciple also went in who came first to the sepulchre : and he saw and believed. For as yet they knew not the Scripture, that He must rise again from the dead."

Summary of the Morrow's Meditation.

We will meditate to-morrow upon two visits to the tomb of Our Lord, as recounted in the gospel of the day, and made, the one by the holy women, the other by St. Peter and St. John. We will then make the resolution: 1st, to bring to the service of God the same fervor as the holy women did when they were seeking the risen Jesus; 2d, to animate ourselves to the practice of virtue by means of the good example set us by our neighbor. Our spiritual nosegay shall be the words of St. Augustine : "*What such and such persons have done, why should I not do?*"

Meditation for the Morning.

Let us adore Jesus, who has risen gloriously from the tomb, allowing Himself to be sought for by the holy women and by the apostles St. Peter and St. John. He acts from love. He hides Himself from the soul in order that it may seek Him; that, in seeking Him, it may desire Him more and more, and that, in desiring Him more and more, it may increase in love and in merits. Oh, how good and amiable is Jesus in all His ways! Let us render Him our homage of adoration, praise, and love!

FIRST POINT.

Visit of the Holy Women to the Sepulchre.

Early in the morning (John xx. 1) of the Sabbath, before day had dawned, the holy women came to the sepulchre of the Saviour, and, finding the stone, which had closed the entrance to it, removed, they ran, filled with grief, to tell the apostles that the body had been taken away. The apostles took these accounts to be only dreams, and would believe nothing. It was an incredulity which entered marvellously into the designs of God ; for it was thereby proved that the witnesses and the preachers of the resurrection did not belong to the number of those credulous minds who believe, without any proof, all that is said to them. They were not only seriously-minded men, who do not believe excepting after strict examination and upon good evidence, but they were also cautious men, who were disposed not to believe mere indifferent proofs, and to yield to nothing except on evidence that was perfectly clear. Now, this was precisely what was necessary, as much to decide the adhesion of the whole universe to the great fact of the resurrection, the basis of the whole of our belief, as to teach us to be neither too credulous nor too incredulous. To believe lightly, and without discernment, is to be imprudent and to be wanting in good sense ; not to believe, because we are determined not to do so, without even being willing to examine whether solid reasons for believing exist, is an infidelity. Wisdom consists in keeping ourselves between the two extremes : to believe nothing lightly, that we may not be deceived ; to lend ourselves willingly to the examination of reasons, with a disposition to believe whatever may be proved. Is this our manner of proceeding? Do we not turn sometimes into ridicule, before having made any examination, the simplicity of those who believe in certain extraordinary facts? Are we ourselves as reserved in our criticisms as we are in our praises? When we study a fact which seems to us strange, do we not do it with prejudice and a desire to find that it is false? Do we bring to this study candor and the love of truth?

SECOND POINT.

The Visit of St. Peter and St. John to the Sepulchre.

Less prompt than the other apostles to condemn the holy women, St. Peter and St. John set out for the sepulchre (John xx. 3). They go there joyfully, because they see, in the absence of the body, the proof that He has risen, according as He had predicted. Faith and love seem to give them wings, and they run with great haste to the sepulchre. Marvellous effect of faith and love! He who believes and loves does everything joyfully; he runs, he flies; nothing stops him; he does not feel the difficulty; he knows nothing of the impossible. With his gaze fixed upon heaven, and with love in his heart, his courage knows no limit. Not only do St. Peter and St. John set out joyfully, but there is between them a holy emulation, which teaches us to rival one another as to who shall be the most fervent, the most humble, the most charitable. St. John arrives the first, doubtless because he was younger, but he does not enter; he remains outside at the door, mortifying in this way his curiosity and, at the same time, yielding to St. Peter the honor of being the first to enter, in order to honor in him the head of the apostolate, the doctor of the faith in whose footsteps all the flock must follow. Peter arrives, sees the linen cloths, with the shroud folded up in a place apart. John comes next; he sees like Peter, and both of them believe without hesitation, not like Mary Magdalene, that their Master had been taken away, but that Jesus had really risen, and therefore was truly God. What a beautiful lesson of mortification, humility, and faith in these holy apostles!

Resolutions and spiritual nosegay as above.

Low Sunday.

The Gospel according to St. John, xx. 19-31.

"Now, when it was late that same day, the first of the week, and the doors were shut where the disciples were gathered together, for fear of the Jews, Jesus came and stood in the midst, and said to them : Peace be to you. And when He had said this He showed them His hands and His side. The disciples, therefore, were glad when they saw the Lord. He said, therefore, to them again : Peace be to you. As the Father hath sent Me I also send you. When He had said this He breathed on them, and He said to them : Receive ye the Holy Ghost. Whose sins you shall forgive, they are forgiven them : and whose sins you shall retain, they are retained. Now Thomas, one of the twelve, who is called Didymus, was not with them when Jesus came. The other disciples, therefore, said to him : We have seen the Lord. But he said to them : Except I shall see in His hands the print of the nails, and put my finger into the print of the nails, and put my hand into His side, I will not believe. And, after eight days, again His disciples were within, and Thomas with them. Jesus cometh, the door being shut, and stood in the midst, and said : Peace be to you. Then He saith to Thomas : Put in thy finger hither, and see My hands, and bring hither Thy hand, and put it into My side ; and be not faithless, but believing. Thomas answered and said to Him : My Lord and my God. Jesus saith : Because thou hast seen Me, Thomas, thou hast believed : blessed are they that have not seen, and have believed. Many other signs also did Jesus in the sight of His disciples, which are not written in this book. But these are written that you may believe that Jesus is the Christ, the Son of God, and that, believing, you may have life in His name."

Summary of the Morrow's Meditation.

We will consider to-morrow in our meditation: 1st, what is the peace which the risen Jesus wishes His apostles every time that He appears in the midst of them; 2d, what is the necessity of this peace; 3d, what is its excellence. We will then make the resolution: 1st, to watch over our interior, so as not to allow it to be invaded by hasty and impetuous movements; 2d, when we perceive that we are in a disturbed state of mind, to pause for a few moments, placing ourselves in the presence of God, in order that we may be re-established in peace. Our spiritual nosegay shall be the words of the Apostle: "*Have peace, and the God of peace and of love shall be with you*" (II. Cor. xiii. 11).

Meditation for the Morning.

Let us transport ourselves in spirit to the cenacle; let us listen reverentially to Jesus saying to His apostles: "*Peace be to you*" (John xx. 19), and let us adore Him with St. Thomas as our Lord and our God (Ibid. 28).

FIRST POINT.

In what the Peace which the Risen Jesus Wished His Apostles Consists.

It consists in the tranquillity of a heart which is always in possession of itself, and which is its own master, without ever being troubled or hurried. It consists in an empire over the passions, the haste, the impetuosity, the excited movements of nature, in order to moderate them, direct them, and prevent them from troubling us. It consists in that sweet liberty of the spirit, which, doing everything at its proper time, with order and wisdom, applies itself to its object without entertaining any regret for the past, without feeling any attachment to the present, without having any anxiety for the future. It consists, lastly, in that calmness of the soul, which, communicating itself to the exterior, impresses on all the actions of the body a certain inexpressible reserve, gentleness, and moderation which is edifying; which is peaceful without being given to slowness; prompt without being hurried; which does not agitate itself, like Martha, with that excessive activity which exhausts the strength, but is tranquil like Mary, listening to Jesus and placing her action in the very repose with which she listens. All its movements are gentle, its operations moderate, its efforts without contention or discomfort; exterior objects do not rouse in it any excited or anxious emotions; or, if sometimes they take it by surprise, it pauses and waits for calmness to return; it is the image of God, who is never troubled any more in the outrages which He receives than in the great works which He performs.

SECOND POINT.

The Necessity of Internal Peace.

Wisdom, says the Holy Ghost, dwells in calmness and repose, not in agitation and tumult (III. Kings xix. 11). I am ready and am not troubled, said David to the Lord, that I may keep Thy commandments (Ps. cxviii. 60). I have held my soul in my hands, so that I may not forget Thy law, he again says (Ibid. 109); thereby signifying that he has freed it from its agitations, and calmed it in its troubles; that otherwise he would have been lost, because trouble is the element of evil, haste the ruin of virtue. The soul which has lost its peace is a prey to all the passions; joy intoxicates and transports it, sorrow casts it down and discourages it; in prayer it is distracted; in recreation it is frivolous; in walking it does not discover either the false steps it makes or the precipices to which it exposes itself; even in the good which it performs it is nature which acts, and not grace. It is incompatible with the Holy Spirit, of which the action, which is always **calm,** cannot harmonize with thoughtless haste, and whose voice cannot be heard in the midst of tumult. And what will become of the soul thus abandoned by its guide and given up to its troubles? If a vessel cannot be guided in a time of calm, who shall answer for it in a time of tempest? Peace of soul is the essential secret and the fundamental stone of the whole interior life. It is the precious peace which must be bought at the cost of all which we possess. The soul which has found it is richer than if it possessed the whole world. Have we until now understood the necessity of internal peace? Do we labor to establish and to keep our soul in this holy state?

THIRD POINT.

The Excellence of Internal Peace.

Internal peace, says St. Paul, passes all understanding (Philipp. iv. 7); and, in fact, it must be something very excellent, since it is the blessing Our Saviour wished His apostles on the eve of His death (John xiv. 1); the blessing which He left them in His testament (Ibid. 27), and which He brings them after His resurrection every time He shows Himself to them (Luke xiv. 36; John xx. 21, 26). Lastly, it is the blessing He charged them to bear everywhere throughout the world (Luke x. 5). This peace is indeed beyond all price; the soul which possesses it hears the slightest sound made by the tempter, and repels it with a strength which is all the greater because it is calm. It observes in its interior all that is not in its proper place, that it may reduce it to order; all that is defective, to correct it; all that is good, to render it better. It has a marvellous facility in prayer, great wisdom in acting, and not less prudence in counselling; in its case, progress in virtue goes on without effort (I. Imit. xx. 6). It fixes itself wholly in the pure love of God, and finds there, as it were, its bed of repose (Aug. *Manual.*, xxix.). All its interior is calm and tranquil; it is like a beautiful sky, in which God delights to make His sun shine; it is like a silent solitude, wherein He loves to speak to the soul; He calls it, and it comes; He draws it, and it runs (Ps. lxxxiv. 9); and it tastes the truth of the words said to St. Arsenius by a heavenly voice: Retreat, silence, and peace. Behold the means whereby to become perfect. Do we employ these means? Do we avoid all that distracts, troubles, and agitates us, and do we apply ourselves to interior and exterior recollection?

Resolutions and spiritual nosegay as above.

Monday after Low Sunday.

Summary of the Morrow's Meditation.

We will meditate to-morrow upon the first obstacle to internal peace, which is excessive activity; and we will consider this activity: 1st, in desires; 2d, in actions. We will then make the resolution: 1st, to combat and moderate our desires; 2d, to do all things calmly and without haste. Our spiritual nosegay shall be the well-known proverb: "*Quick enough, if it is good.*"

Meditation for the Morning.

Let us adore the profound peace which God has enjoyed from all eternity, amidst an infinite number of occupations, and which, at the same time, He makes His saints enjoy. For peace is the true blessing of Heaven, the portion of the blessed. Let our hearts overflow with praise to God and His saints, and let us ask for ourselves the reformation of that excessive activity which is the greatest obstacle to peace.

FIRST POINT.

Of Activity in Desires.

There is in us a tendency to throw ourselves with passionate ardor into everything which promises us some fitful glimmers of pleasure, even if it be only a pleasure arising from curiosity. It is a consequence of our nature, which has a consciousness of being born for infinite happiness, and which, although it well knows that it will not find it anywhere here below, is sometimes weak enough to seek for it. Hence activity in desires; and if we do not put a stop to the impetuosity of this first movement, the heart is taken captive and attaches itself, troubles and agitates itself. Carried away by the desires which do infinite harm to the soul, they blind it; they make it forget God, and, at the same time, the views of faith which ought always to direct it. They accustom it to seek itself in everything; and, in the preoccupation into which they cast it, it thinks of nothing but gratifying itself. The fear of not succeeding embitters and angers it; and if, in the end, it does not succeed, it murmurs and is irritated; it indulges in ill-tempered words, and is an annoyance to others and to itself. Hence the prayer in Ecclesiasticus: "*Lord, turn away from me all coveting*" (Ecclus. xxii. 5); and that of David: "*Give me not up, O Lord, from my desire to the wicked*" (Ps. cxxxix. 9). Hence, also, the counsel of the saints, that we must never act instantly upon the impulse of a too-hasty desire, but we must know how to wait until the agitation has passed away and calmness and composure have returned, that we may examine the desire as to whether it is good; if it will not be agreeable to God, that we should renounce it; and even if we feel, after examination, that the desire is legitimate, not to act in virtue of the desire, but only from a wish to act in accordance with the good pleasure of God. We must never, say the saints, wish to be anything but what God wills we should be; we must never, in temptation, desire tranquillity, repose in the midst of affairs, light in darkness, consolation in dryness, solitude in company, company in solitude. Even desires for better things cease to be good as soon as they are not sufficiently moderate to be swallowed up in the sole good pleasure of God. This is what made St. Francis de Sales utter these sweet words: "*I desire but few things, and the little I desire I desire very little; and if I could be born again, I should wish not to have one single desire.*" And St. Aloysius uttered these other words: "*I banish from my heart, not only desires for indifferent things, but also desires for even holy things, when they are too eager.*" Let us examine ourselves before God where we stand in regard to the practice of these holy rules.

SECOND POINT.

Of Activity in our Actions.

In order to live in peace, we must always act

with simplicity and moderation; not infuse passion into anything, but keep our soul always in a state of equanimity, energetically restraining the vivacity and the impetuosity of nature and the boiling agitations of a heart which does not possess itself. There are certain minds, and perhaps we are of the number, who can do nothing deliberately and calmly, who are always on the run, and who do not know how to walk. Before they act, they are preoccupied, forestalling the hour of action in their thoughts, contrary to the advice of the sage: "*Everything in its season*" (Eccles. viii. 6), and to the example of Our Lord, who said: "*My hour is not yet come*" (John ii. 4). He did not hasten the hour, He did not delay it; He waited for it peacefully. During action, hardly have these men who are devoured by activity begun, before they long to have finished; they hurry, they hasten, without having regard to the proverb: "*Quick enough, if it is good.*" Thenceforth there is no more peace, no more thought of God, or, at any rate, they seek themselves with God, and they place themselves beside Him; henceforth there is distraction in the soul, which is entirely given up to the exterior portion of its action; there is often even scandal given, from want of order and of a rule in the manner of acting. The soul comes into collision with the obstacles it meets with, and injures itself in consequence; it is distressed and impatient. It was this state which St. Bernard deplored in himself when he said: *"There is nothing tranquil in me."* Happy he who knows how to moderate his vivacities, always to possess himself, and to be master of himself.

Resolutions and spiritual nosegay as above.

Tuesday after Low Sunday.

Summary of the Morrow's Meditation.

We will meditate to-morrow upon two other obstacles to interior peace; that is to say: 1st, preoccupation with affairs; 2d, discouragement after faults. We will then make the resolution: 1st, often to examine ourselves, in the midst of our occupations, in order to make our soul resume the calm of peace beneath the eye of God; 2d, never to be discouraged after our faults. We will retain as our spiritual nosegay the words of Our Lord to His apostles: *"Let not your heart be troubled"* (John xiv. 1).

Meditation for the Morning.

Let us transport ourselves in spirit to the cenacle, into the midst of the assembly of the apostles; let us fall on our knees at the feet of the risen Jesus who brings to them peace. *"Peace be to you,"* He says to them. Let us beg of Him to say the same words to our hearts, and let us render Him our homage of adoration, of praise, and of love.

FIRST POINT.

To be Preoccupied with Business Occupations an Obstacle to Interior Peace.

To maintain interior peace is, it seems, an easy thing for a religious living in his cell, a stranger to the world and its concerns, to its affairs, and to the thousand occupations which sometimes absorb every moment of life, but to maintain interior peace when, from morning until evening, we are besieged by occupations, which, like a troubled sea, invade us and do not leave us any repose (Is. lvii. 20), is quite a different thing; this is what some persons even declare to be impossible, but it is an error. In order not to lose interior peace in these positions, it is sufficient to take each affair, one after the other, with the same calmness and the same freedom of mind as though we had until then done nothing, and as though we had nothing to do afterwards, with a complete absence of all anxiety and of all agitation, for the very evident reason that the more we have to do, the more necessary it is not to be disquieted; that it is in the tranquil and sedate mind which is in full possession of itself that the wisdom is found which performs everything in the manner it should be done; that even when anxiety and eagerness may be able to effect something which is good, it is not the less necessary to maintain peace in the soul; because the whole universe is not worth interior peace; because, also, after sin there is no greater evil than disquietude; because, lastly, in this eager movement, in this kind of interior fermentation, there is the most frequently only what is natural and human; grace counts for nothing in it; the Spirit of God is not there, and nothing wise or of what is sufficiently reflected upon is possible under the empire of preoccupation. It was thus that St. Vincent de Paul understood the matter, when, amidst his most absorbing occupations, he was always so calm, so entirely master of himself, as was attested by the habitual tranquillity of his soul and of his countenance. Do we follow these rules of conduct?

SECOND POINT.

Discouragement after Faults another Obstacle to Interior Peace.

Often, when viewing its miseries and weaknesses, the soul is troubled and discouraged, sometimes by a deep feeling of its powerlessness to do what is right, sometimes through self-love indulging in a fit of spite. What! after so many resolutions always to be falling; after so many remedies to be always ill; after so many prayers always to have the mind so distracted, so frivolous, and the heart so cold! Who is there that would not lose peace? It is in this way we sometimes reason, but David says: It is in vain that man is disquieted (Ps. xxxviii. 12); discouragement and the disquietude which it produces never have done any good or cured any kind of evil. Disquietude obscures reason, throws the interior into a state of disorder, lowers courage, makes the will like ice, and renders the mind incapable of receiving divine light. It is much better, therefore, to humble ourselves peaceably before God and, confessing our profound wretchedness, acknowledge that without His succor we should have done worse still, and thank Him with our whole hearts that we have not done worse still, since there is no kind of sin of which we are not capable, if the grace of God does not restrain us. After this humble confession, we rise again full of confidence in the divine mercy, we cast ourselves into the arms of God, with a heart full of love, like the prodigal son into the arms of his father, and we encourage ourselves to repair the faults we have committed by leading a better life. As soon as we can, we make a good confession, and we think no more of our sins, at any rate in detail, because the remembrance of them would give the devil an opportunity of either tempting us again or of troubling us with the fear of not having declared all the circumstances; we must only retain a confused and general idea of our sins, to keep us in an humble state of mind.

Resolutions and spiritual nosegay as above.

Wednesday after Low Sunday.

Summary of the Morrow's Meditation.

We will meditate to-morrow upon two other obstacles to the peace of the soul, that is to say, vain joy and a wrong kind of sadness. We will then make the resolution: 1st, not to yield to joy which carries away and dissipates the soul, but to moderate it by means of some moments of reflection before God; 2d, in fits of sadness to raise our courage by confidence in God and the hope of heaven. Our spiritual nosegay shall be the words of the Blessed Virgin: *"My spirit hath rejoiced in God my Saviour"* (Luke i. 47).

Meditation for the Morning.

Let us adore Jesus Christ bringing us peace from the day of His birth as the first pledge of His love (Luke ii. 14), and offering it to us again after His resurrection as the first fruits of all His merits. Let us thank His divine goodness for so inexpressible a blessing, and let us promise Him to remove the obstacles which prevent us from enjoying it.

FIRST POINT.

Vain Joy an Obstacle to Interior Peace.

Doubtless there is a good and praiseworthy joy which is a gift of Heaven, a fruit of peace with God, a solace in the miseries of this life, even a help to virtue, which easily flags in sadness, one of the charms of religion, of which it reveals the heavenly amiability to the world, lastly, a need which man feels and without which he would easily become discouraged in the practice of virtue. It is that joy which St. Paul calls a fruit of the Holy Spirit (Gal. v. 22), and which he so strongly recommends to the faithful (Philipp. iv. 4). But outside this holy joy, which has its principle in God, there is another kind of joy, which has its root in the creature; and it is that which we say is incompatible with interior peace. It is a perfidious joy, which we do not distrust because of its only producing pleasure, and a pleasure which is often not criminal; because, really, the first of the wounds which it inflicts on the soul is inattention to ourselves, which prevents us from feeling the other wounds as well as the first which it inflicts. This inconsiderate joy deranges the whole economy of the interior, dissipates it within, attracts it without, makes it speak and act without reflection; it is that of which the Holy Spirit says: *"The heart of fools is where there is mirth"* (Eccles. vii. 5); it is an enemy of restraint and mortification; it makes a man speak loud, burst into laughter, forget the rules of modesty in his demeanor, his gestures, his looks, his manners; it delivers up the soul to all the flights of imagination, opens the door of the senses to exterior objects, by the aid of which it puts into movement all that is within us, and excites there a tumult which agitates us and upsets the heart; and in this kind of interior ebullition all the unction of piety evaporates; a quarter of an hour of this joy, says the Imitation, scatters the fruit of several days of recollection (II. Imit. xxi. 11); often a long period and many efforts are required to make us regain what we have lost. Do we not often allow ourselves to indulge in this vain joy which is the ruin of piety?

SECOND POINT.

A Wrong Kind of Sadness an Obstacle to Interior Peace.

Sombre sadness often succeeds vain joy, says Holy Writ (Prov. xiv. 13), and peace receives as much injury from the one as from the other. This sombre sadness concentrates us on ourselves, makes us discontented, irritable, impatient, disagreeable to others and to ourselves, and there is in it a source of many evils (Ecclus. xxv. 17). If then we wish to establish and maintain in us the reign so desirable of peace, we must keep ourselves from black sadness as well as from excessive joy, and maintain between these two extremities the just medium of a tranquil and moderate joy. When we are inclined to sorrow we must recall to mind the joy of Mary when she visited Elizabeth, that joy in God so greatly recommended by the Holy Spirit (Ps. xcix. 2; Ibid. xxxvi. 4), so much preached by the Apostle (Philipp. iv. 4), and which is derived from the consideration of the infinite goodness of God, of His tenderness, of His mercy, of His love, of the heaven He commands us to hope for, and of the magnificent recompenses which He reserves for us there. When, on the contrary, fits of vain joy threaten to carry us away, we must recall the sadness of Mary at the foot of the cross; a sadness according to God, which the Apostle praised in the Corinthians (II. Cor. vii. 9), and with which we ought to be inspired by the consideration of Jesus Christ suffering and dying for us, the sight of all the evils with which religion and the Church are afflicted, the thought of the souls which are lost, the remembrance of our innumerable faults, the consciousness we have of our habitual sins; for, says the Imitation, we often allow ourselves to indulge in vain joy when we ought to weep (I. Imit. xxi. 2). In the evil days wherein we are, what matter for serious reflection, for just grief, which ought to make us pray and sigh, without, however, giving way to the sombre melancholy which takes from us the peace of our souls.

Resolutions and spiritual nosegay as above.

Thursday after Low Sunday.

Summary of the Morrow's Meditation.

We will meditate to-morrow upon two other obstacles to interior peace, that is to say: 1st, temptations; 2d, scruples. We will then make the resolution: 1st, to turn away peacefully from our temptations at the very first moment we perceive them; 2d, to serve God with a heart at ease, with confidence and love, without tormenting ourselves with the fear of displeasing Him. Our spiritual nosegay shall be the words of the Lord's Prayer: *"Lead us not into temptation, but deliver us from evil."*

Meditation for the Morning.

Let us adore Jesus Christ giving us His peace, in the person of the apostles (John xiv. 27). Let us thank Him for so precious a gift, and let us ask of Him grace not to allow it to be taken away from us by temptations or scruples.

FIRST POINT.

Temptations an Obstacle to Interior Peace.

Wherefore in temptation art thou sad, O my soul, and wherefore art thou troubled? (Ps. xli. 6, 12.) Let us in the presence of God examine whether we have any cause for losing peace in these circumstances. Shall it be because we believe that temptation is in itself a sin? But Jesus Christ and all the saints also were tempted, and no bad thought, no imagination, even the most hideous, is in itself a sin,—all these thoughts and imaginations are constantly present to the eyes of God, who sees all things, and they do not in the very slightest degree tarnish His infinite purity. Shall it be that we are afraid of having consented to the temptation? But even if we should have consented, we must not allow ourselves to be troubled, since we have seen in the previous meditation that even after our faults we must preserve peace, and that to lose it would be a sin added to another sin. Then if the temptation has displeased us, molested and saddened us, if it has nevertheless inspired us with horror, if it has been only in spite of ourselves and against our will that we have submitted to it, we have in that the very proof itself that we have not consented to it. We sin only with our will; what is against our will cannot be imputed to us. Shall it be that we are afraid of consenting later on? But why lose confidence in God, and not hope that He will sustain us if we pray earnestly to Him, if we mistrust ourselves, if we avoid occasions, and do not presume on our own strength? Shall it be, lastly, that this life of combat and of strife wearies us? But, 1st, we can diminish these combats and these strifes by despising the tempter to the extent of not even deigning to give him a thought in order to answer him, to turn our backs upon him instead of fighting with him, like the woman who hears a dog bark at the door without taking any notice of it, and continuing to occupy herself peaceably with what she is doing. Like her let us permit the devil to bark outside without paying any attention to him, and let us continue in peace the work we have to do. 2d. We may diminish our temptations by not reflecting upon them when they have passed away to see whether we have consented to them, because to reflect upon them would be the means of making them revive; we ought not to remember them excepting as a whole, in order to arouse in us vigilance and the spirit of prayer, and to abase ourselves before God in the sentiment of our own misery, like St. Teresa, who said: "O God, how worthless I am! Behold, this is what my evil nature can produce; behold, this is indeed the fruit of my garden;" and on the other side in admiration and love of the divine goodness : "O God, how good Thou art to abase Thy love to me." Is it thus that we conduct ourselves in temptations?

SECOND POINT.

Scruples an Obstacle to Interior Peace.

There are scrupulous persons who are tormented by the fear of not loving God enough; they are anxious, weary their brains, exhaust themselves by efforts and discussions; they put their hearts under a press to force affections to issue from it, and they expose their mind to torture that they may put a term to its inconsistencies. Ah, this is not according to God. God wills that all should be gentle and moderate in His service; He only asks of us a solid preference, a will which is resolutely determined to conform our will to His. When we have told a friend that we love him sincerely, that all we have is at his disposition, we think that he ought to be glad, although we have said it all simply and without effort; it is the same with God.

There are other scrupulous persons who are tormented by the fear of not being in a state of grace and of sinning in all that they do as well as in all that they say. There are three remedies for this evil: 1st, to obey the confessor. We always do well when we obey, and even if the confessor were mistaken, obedience would excuse the penitent in the sight of God; 2d, in a state of doubt the scrupulous person ought never to believe that he has committed a mortal sin; he ought always to pass sentence in his own favor, unless when he has a certainty in regard to his culpability which excludes all doubt; 3d, we must always look upon God as a good Father, who desires to see in those who serve Him simplicity, confidence, the love of a child, and not the fear of a slave, which is uneasy, torments itself, repeats its prayers incessantly, and would like to repeat its confessions. Observing these rules, we shall always enjoy peace, in spite of scruples.

Resolutions and spiritual nosegay as above.

Friday after Low Sunday.

Summary of the Morrow's Meditation.

After having meditated upon the obstacles to interior peace, we will meditate upon two means of establishing this peace in our hearts, that is to say: 1st, to combat thoughts consisting of self-love which trouble us with humble sentiments of ourselves; 2d, to renounce all the effeminacies belonging to a sensual life, which preoccupy the soul and lead us to seek ourselves in everything. Our spiritual nosegay shall be the words of God to Isaias: "*I will bring upon her* [*the humble and faithful soul*] *as it were a river of peace*" (Is. lxvi. 12).

Meditation for the Morning.

Let us adore Jesus Christ canonizing men of peace, in order to invite us all to partake of this happy disposition. Let us render to Him, on this account, our homage of gratitude and love; let us beg Him to give us grace to obtain this peace by the double means of humility and renunciation of a sensual life.

FIRST POINT.

Humility a Means for Obtaining Interior Peace.

The truly humble soul is always calm and tranquil. Praises surprise it instead of elevating it, because it esteems itself to be unworthy of them; blame rejoices it instead of casting it down, because it is a pleasure to it that others should think of it as it thinks of itself. Calumny does not disconcert it, because it feels that if it is wrongly reproached with certain faults, there are others for which it might with justice be blamed, and that, having formerly received praises which it did not merit, it is just it should now receive reproaches which it does not deserve. Wealth and prosperity do not make its heart swell; it receives them modestly, because it esteems itself to be unworthy of them; adversities and trials do not cast down its courage; it receives them with meekness, believing that it merits them. Preferences shown to others do not afflict it; it is always raised too high, in its own opinion, as long as there remains a place below it; and, even if it were relegated to the lowest rank, it would believe itself to be only too highly favored to be permitted to occupy it. The wicked may vomit out all their venom against it; far from being soured by their malice, it praises them for their discernment, and unites itself sincerely with them to undeceive those who esteem it beyond what it believes itself to deserve; and thus, whatever may be done to it, whatever may be said or thought of it, it always enjoys perfect repose in the deep conviction of its own nothingness, and is filled with a continual serenity; so true are the words of Jesus Christ: *"Learn of Me, because I am meek and humble of heart, and you shall find rest to your souls"* (Matt. xi. 29), and those words also of the Imitation: *"The more humble we are, the more we are at peace"* (I. Imit. iv. 2). Oh, it is very different with the heart which is not humble! It is always troubled and sorrowful. A preference accorded to another grieves it; the reputation attained by its neighbor embitters it; a want of consideration casts it down; a slight humiliation confounds it; an imaginary show of contempt disconcerts it; the mere fear of sinking in the esteem of others tears its soul to pieces; and anxiety as to what is said or thought of it overwhelms it; even praises and honors do not less injure its peace; it first experiences a false joy, which dazzles it; then, a few moments afterward, an inexplicable feeling of discontent and remorse. Let us question our conscience, and see whether all this is not true.

SECOND POINT.

Renunciation of a Sensual Life another Means for Obtaining Interior Peace.

The sensual life and interior peace are like two enemies in face of each other, of which the one fortifies himself with what weakens the other; because: 1st, the more we grant to the senses, the more they demand, and they weary the soul by ceaseless requirements. *" The eye is not filled with seeing, neither is the ear filled with hearing"* (Eccles. i. 8). Never do gluttony and effeminacy say: It is enough; and in proportion as their demands are gratified they become sensitive and difficult to satisfy. Nothing is good enough for them; nothing pleases them; the least contrariety makes them revolt; everything troubles the sensual soul and puts it into a bad temper. 2d. The satisfaction of the senses attracts the soul to outside things and occupies it with exterior objects, which distract it and render it incapable of the things of God, above all, of interior peace; whilst the man who has renounced a sensual life remains as calm in his interior as though he were in an impregnable fortress, disdaining the cries of the senses, which call him to enjoyment, and thus forming in his soul a robust and vigorous temperament, in the same way as the body is rendered healthier and stronger by fatigue and labor. Let us here examine ourselves, and let us consider how the love of enjoyment renders us frivolous and gives up our soul to exterior things.

Resolutions and spiritual nosegay as above.

Saturday after Low Sunday.

Summary of the Morrow's Meditation.

We will meditate to-morrow upon another means of acquiring interior peace, which is the perfect conformity of our will to the will of God; and, in order to understand it, we shall see: 1st, that no peace is possible with attachment to self-will; 2d, that perfect conformity to the will of God gives a delicious peace. We will then make the resolution: 1st, not to desire any other talents, or any other condition, or any other fortune than what God has given us; 2d, lovingly to follow, in all the details of our life, the will of God, even as the Magi followed the star which led them to Bethlehem. For our spiritual nosegay we will often say to God: "*Father, Thy will be done.*"

Meditation for the Morning.

Let us return to the cenacle, and let us listen to Our Lord repeating the words which do such good to the soul: "*Peace be with you.*" Let us ask of Him grace to purchase the happiness of interior peace at the cost of our own will, by putting in its place the most adorable, most holy, and most amiable will of God.

FIRST POINT.

There is no Interior Peace where there Exists Attachment to Self-will.

Whoever has an attachment to his own will condemns himself to trouble and unhappiness. Sometimes this will comes in contact with the will of another and then it is hurt or painfully wounded, or the heart is embittered and rendered uneasy; sometimes it comes in contact with itself—desiring one thing one moment, and the contrary another; at another it pursues what it cannot reach, or, if it ends by reaching it, it quickly feels the emptiness of it and is disgusted. Thus dragged from one side to another, and torn by a thousand ceaselessly-renewed desires, it is always discontented with itself or with others; it never follows its own will without being made sorrowful. It is indignant if others oppose it; it is grieved if it is obliged to renounce its desires, or if it can satisfy them it reproaches itself with having yielded to passion rather than to reason, and the result is a discontent with itself which is the great enemy of interior peace. O Christian soul, says the author of the Imitation, why dost thou render thyself unhappy? If thy will desires to satisfy itself here below, thou wilt never be at peace, never without trouble or anxiety, because everywhere thou wilt meet with contradictions, everywhere something will be wanting to thy happiness (III. Imit. xxvii. 2).

SECOND POINT.

Perfect Conformity to the Will of God gives a Delicious Peace to the Soul.

Nothing in this world can trouble the peace of him who wills only the will of God. In everything which happens, on the part of men or in events, he reveres the Divine Will which directs all things; and this keeps him in an unchangeable serenity of soul, which the emotions of the passions or the winds of desires cannot trouble. Lord, said David, I have seen in all things Thy good pleasure, which has guided events, and has led me as by the hand (Ps. lxxii. 24). What delicious peace is it which we enjoy beneath the guidance of so amiable a hand; above all when we mingle our self-love with it only for the purpose of uniting it with that of God like the holy king, who exclaimed: *"What have I in heaven? and besides Thee what do I desire on earth?"* (Ps. lxxii. 25.) In this happy state we may see everything changed and overthrown around us, but we are still in peace, because, on the one hand, we know that nothing happens except by the command or the permission of God, and, on the other hand, we desire with our whole will all that Providence ordains or permits; nothing more, nothing less, nothing otherwise. It is even true to say that we are never put out, that we have always everything that we wish for, that we suffer only what we will to suffer, because we wish and we will nothing but the will of God, who governs and disposes all things. Then the soul grows and raises itself above the tempests and agitations of the world into a higher region, a region of peace and of serenity, whence we soar above all the storms which rage here below, and which we no longer hear roaring except far beneath our feet; a region of ineffable calm where the soul reposes, deliciously overwhelmed with love of the Divine Will (II. Imit. xxxi. 1). Then if the tongues or the malice of men shoot their arrows at us and wound us, we receive the shots, not as coming from the inimical hand which has directed them, but as coming from the paternal will of God, which only makes useful wounds, and can cure them when it so pleases Him. Then when prosperity arrives we receive it not with that intoxicated joy which troubles interior peace, but with a kind of modest fear because we appreciate the danger of it; if adversities visit us, we receive them, if not with the joy of the primitive Christians, which would be far better, but at least with resignation and confidence in Providence, like the holy man Job (Job i. 21) ; and in the one case as in the other we say to God : *"My heart is ready, Lord, to receive from Thy hand adversity as well as prosperity"* (Ps. cvii. 2 ; St. Augustine). Then, lastly, whatever happens, the soul remains calm and at peace, because it looks upon itself as being always under the eye of God, which sees all, under His power, which can do all, under His action, which concurs in all, or rather in the arms of His love, which wills all that is good for us.

Resolutions and spiritual nosegay as above.

SAINTS

WHOSE FEASTS, BEING ON FIXED DAYS, DO NOT FOLLOW THE VARIABLE COURSE OF THE LITURGY.

February Twenty-fourth.—St. Matthias, Apostle.

Summary of the Morrow's Meditation.

We will to-morrow consider in our meditation: 1st, what God did to St. Matthias in his election to the apostolate; 2d, what St. Matthias and the apostles who elected him did for God. We will then make the resolution: 1st, to thank God every day for His benefits, and to correspond with them, as this apostle did, by a holy life; 2d, to have in all things a very pure intention to please God. We will retain as our spiritual nosegay the words of St. Paul to the Ephesians: "*Blessed be the God and Father of Our Lord Jesus Christ, who hath blessed us with every spiritual blessing in heavenly places, in Christ, as He chose us in Him before the foundation of the world, that we should be holy and unspotted in His sight in charity*" (Eph. i. 3, 4).

Meditation for the Morning.

Let us adore God calling St. Matthias to the

apostolate, that he might replace the traitor Judas; let us bless Him for this choice, which was so happy a one for the Church, and with this object in view let us render Him all our homage.

FIRST POINT.

What God did for St. Matthias in his Election to the Apostolate.

The apostles had chosen two candidates for the place which was vacant in the apostolic college, Joseph, surnamed the Just on account of his eminent sanctity, and Matthias. Spite of the renown which surrounded Joseph, God, speaking by the voice of lots, designated Matthias: a palpable proof that He does not judge like men or that He gives the preference to whosoever pleases Him even in spite sometimes of the want of merit. He is the master of His gifts, and He disposes of them according to His pleasure, without our having the right to ask of Him the reason (Rom. ix. 16, 18). In right of this liberty which is essential to Him, He chooses St. Matthias. Honor, praise, and love be to His infinite mercies! Jesus Christ might Himself have chosen before His ascension this happy privileged man; He preferred to leave the choice to the drawing of lots: 1st, in order to show forth the truth that He governs human things from the heights of heaven; 2d, in order to authorize the supernatural economy of His Church and to show that He presides invisibly over it by a special Providence ; 3d, to keep both the electors and the elected humble, as well as the superiors whom He charges to govern in His place and the inferiors who are subject to their guidance ; humility being necessary to the first in order to command with meekness, to the second to obey with love, to both the one and the other in order to imitate Him who, being the greatest of all, made Himself the least, that He might make them sharers in His greatness. It is thus that the goodness of God shows itself in the election of St. Matthias. God is good in all that He does. Let us admire His goodness towards us, to whom He sends His graces, anteriorly to all our merits, through the pure and simple mercy which anticipates all our needs.

SECOND POINT.

What St. Matthias and the Apostles who elected him did for God.

1st. The apostles, in this election, had before their eyes no other interest than that of the glory of God ; no other aid than His goodness; and they asked of Him, with fervent prayers, to show them him whom He had chosen in His eternal designs (Acts i. 24). "Thou who knowest the hearts of all men (Ibid.), show to us him whom Thou dost prefer for this ministry of which Judas deprived himself by his crime." A beautiful lesson for us, and one which teaches us in all things to consider God and His good pleasure, without respect to whatever may be the opinion of men. 2d. St. Matthias being elected, corresponds to the grace of his election by a wholly apostolic zeal. Judea falls to him as his share in the distribution of the provinces. He signalizes himself there by converting a great number of the heathens. From thence, carried away by his zeal, for which such narrow bounds do not suffice, he advances as far as Ethiopia; and there makes war upon error and ignorance. His conquests irritate his enemies; they put him to death and crown his apostolate with the glory of martyrdom. Let us learn from this always worthily to fill our position, whatever it may be, and to excel in all the labors which Providence confides to us, according to the precept of the sage: *"In all thy works keep the pre-eminence"* (Ecclus. xxxiii. 23).

Resolutions and spiritual nosegay as above.

March Nineteenth.—St. Joseph.

(FIRST MEDITATION.)

Summary of the Morrow's Meditation.

We will to-morrow consider the two fundamental titles of the devotion to St. Joseph: 1st, he was the father of Jesus; 2d, he was the spouse of Mary. We will then make the resolution: 1st, at this feast to renew our devotion to St. Joseph; 2d, often and confidently to invoke him. We will retain as our spiritual nosegay the words of Pharao respecting Joseph the son of Jacob: "*Go to Joseph.*"

Meditation for the Morning.

Let us raise ourselves in spirit to heaven before the throne of St. Joseph; let us thank God for the greatness to which He has raised him, and let us lay at the feet of this holy patriarch all that our hearts contain of praise, of admiration, and of love.

FIRST POINT.

St. Joseph, the Father of Jesus.

How different are the judgments of God from those of men! If the world had been consulted upon the choice of him to whom was to be confided the adoptive paternity and the care of the Word Incarnate, it would doubtless have fixed upon some powerful monarch; but what, in the presence of God, are all earthly grandeurs? With a breath He casts them down and makes them disappear like the dust before the wind. Perhaps it would have advised that he should have been some rich person, who could have brought up the Child amidst all the refinements of luxury; but what, in the sight of God, are all riches except as so much dust, all the comforts of life and the pleasures of the senses except effeminacy, unworthy of a just heart? For want of finding a man worthy to fill such a position, perhaps the world would have raised its thoughts to heaven, and have imagined that it would be necessary to depute some prince of the heavenly court to come here below and surround, with his care and his love, the Incarnate God. But no; even the highest angels are not judged to be worthy of so august a mission. Greater, in the Divine estimation, than all the principalities of heaven and of earth, the humble and just St. Joseph was the elect of God, destined to receive within his arms Him who, from all eternity, dwelt in the bosom of the Father; to lodge in his house Him whose palace is the heavens; to be, lastly, the father of Jesus, as the Holy Spirit calls him in the gospel (Luke ii. 33), and Mary in the temple (Ibid. 48). In virtue of this choice, Joseph was substituted for God Himself, commissioned to hold His place in the eyes of men, honored with a paternity of jurisdiction over Jesus, in virtue of which he had a right to command the Incarnate Word to perform all the duties of a son, and the Word, by whom all things are made, obeyed him as though he were His father. O sublime title! O incomparable dignity! It was through this title that on the day of the Circumcision Joseph gave to the Infant God the name of Jesus; that on the day of the Presentation He took the Ruler into the temple and placed Him upon the altar; that during the days of His infancy he received His sighs, appeased His cries, dried His tears, solaced His pains, and that, in order to shelter Him from the fury of Herod, his paternal hands carried Him into Egypt, far more honored thereby than if they had borne the sceptre for which they were made. It was through this title, lastly, that he fed Him at the cost of his labors and his sweat; that he gave Him lodging and clothing, and provided for all His wants. What a heavenly and magnificent mission! How it aggrandizes St. Joseph! How it ought to render the devotion to him dear and venerable to us! How it ought to inspire us with confidence in his protection!

SECOND POINT.

St. Joseph, Spouse of Mary.

Under this new title, Joseph shares the glory of the Blessed Virgin, chosen from amongst all the descendants of Juda, and raised above all nations, who will forever call her blessed. As being the husband of Mary, he is her master and lord; for the Apostle says that "man is the head of the woman." So that, O miracle of elevation! the Mother of God, the Queen of heaven and earth, calls Joseph her lord and master, and she from whom the angels consider it an honor to receive commands was subject to him in all things. As the husband of Mary, he was her guardian angel, called to the honor of being her conductor in her journeys, to console her in her afflictions, to shelter, during the season she was with child, from calumnies that living tabernacle filled with the glory of the Most High. As husband of Mary, he had a portion of holiness proportioned to the holiness of the heavenly virgin. For, if it be right that the two persons united together should each bring as their dowry an equal share of possessions, who can doubt that Heaven, the author of this marriage, should not have arranged, in order to render it suitable, a certain proportion between the merits of Joseph and those of his holy spouse?

O Joseph! at one and the same time father of Jesus and spouse of Mary, how great, then, art thou! What a beautiful place thou must occupy in heaven! From here below I lovingly contemplate thee, raised upon a throne which very far surpasses the twelve thrones on which the apostles, who are destined to judge the twelve tribes of Israel, are seated. For the apostles are only the servants of Jesus Christ; and He calls thee His father. It seems to me that I behold thee seated at the very side of the Queen of heaven, since thou art her husband. By what profound enough homage can we ever sufficiently honor thee, venerate thy statues and thy pictures, celebrate thy feasts and invoke thy name, make of it our delight and often repeat it!

Resolutions and spiritual nosegay as above.

March Twentieth.—St. Joseph.

(SECOND MEDITATION.)

Summary of the Morrow's Meditation.

We will meditate to-morrow upon the confidence which we ought to have in St. Joseph; and we shall see that this saint unites eminently in his person two qualities which command confidence; that is to say : 1st, power, or sufficient credit to obtain what we ask of him ; 2d, goodness, or a heart benevolent enough to charge himself with our requests. We will then make the resolution : 1st, to have recourse to Joseph in all our troubles and difficulties ; 2d, to have a special devotion to him, loving his altars, his pictures, his feasts, the prayers composed in his honor, and cherishing his memory. Our spiritual nosegay shall be the same as yesterday : *"Go to Joseph."*

Meditation for the Morning.

Let us adore the goodness of God, who has established St. Joseph in His Church, there to be the hope of Christians, the consolation of the afflicted, the saint to whom all who have graces to solicit may have recourse. Let us thank Him for His divine goodness, and let us congratulate St. Joseph on the glorious mission confided to him by Heaven.

FIRST POINT.

The Power of St. Joseph to Obtain all that we ask of him.

The power of St. Joseph greatly surpasses the power of all the saints and angels put together. For at one and the same time he has power over the heart of God, power over the heart of Jesus, power over the heart of Mary. What could God refuse to a saint whom He has chosen in preference to all the princes of heaven and earth, whom He has associated with His divine paternity, whom He has constituted the visible providence of His Son, and who so worthily fulfilled that divine mission? What could the Incarnate Word refuse to him from whom He had received everything on earth; who had provided Him at the cost of his sweat with all things necessary to life; who had labored and lived for Him alone; to him who had loved Him so much, served Him so well, who had watched over Him so carefully and vigilantly? What could He refuse to him whom He had so loved when He was on earth that He obeyed his slightest desires, as though they had been express orders? Could it be that in heaven His feeling should have undergone a change with regard to His adopted father? Would it be possible for Him to be no longer grateful for his kindness or sensible to his wishes? It is a thing not to be supposed; and even if Joseph could not of himself obtain a favorable answer to his requests, has he not over the heart of Mary, in order to make her support his requests, the most undeniable of rights, the rights of legitimate authority, the rights of gratitude and those of friendship? Oh, how Mary would cast herself at the feet of Jesus rather than see her holy spouse refused! Now if Mary and Joseph are for us, who shall be against us? We can therefore say of Joseph what the Fathers have said of Mary: that her prayer is all-powerful to obtain what she asks for.

SECOND POINT.

The Goodness of St. Joseph, or his Kind Readiness to Employ his Credit in our Favor.

How good must he be, he whom the hand of God and the grace of the Holy Ghost have formed to be the father of Jesus and the spouse of Mary; he who so often bore in his arms and laid to rest on his bosom the charity of the Incarnate God! Oh, how measureless in degree and how overflowing must have been the charity which filled the one heart and the other! If, through having reposed during a few moments upon the bosom of Jesus, St. John became the apostle and the evangelist of charity, what must St. Joseph have been after living for so many years in intimate intercourse with Jesus and Mary? What love, what tenderness, what a disposition to render services, to accept all the requests made to him and to obtain a favorable answer to them, must he not have derived from this double source of goodness! Let us conclude from all this how good St. Joseph is. His goodness equals his power, and experience proves it. "I do not remember," says St. Teresa, "ever to have asked anything from God through the intercession of St. Joseph that I have not obtained it; and I do not know any one," she adds, "who has assiduously invoked him without making notable progress in virtue." Several other saints attest the same thing by their own experience, as well as by the experience of others, and thence infer that the devotion to St. Joseph is one of the surest marks of predestination. Let us ourselves hence conclude how much we ought to love St. Joseph, to cherish his remembrance, and to pray to him with confidence.

Resolutions and spiritual nosegay as above.

March Twenty-first.—St. Joseph.

(THIRD MEDITATION.)

Summary of the Morrow's Meditation.

We will meditate to-morrow upon the special graces attached to the devotion to St. Joseph; and taking our stand upon the pious belief of the Church, that God, in order to recompense the saints who have excelled in a certain grace or a virtue, grants them in heaven a special power to obtain for their clients the same grace or the same virtue, we shall see that to the devotion to St. Joseph are attached, on this account, four principal graces: 1st, a grace of love towards Jesus and Mary; 2d, a grace of purity and innocence; 3d, a grace of interior life; 4th, the grace of a holy death. We will then make the resolution: 1st, often to ask, through St. Joseph, for the holy graces in which he himself so greatly excelled; 2d, to exercise ourselves, by means of the example of St. Joseph, in the practice of the love of Jesus and Mary, of purity, and of the interior life. Our spiritual nosegay shall once more be, "*Go to Joseph.*"

Meditation for the Morning.

Let us adore the Spirit of God communicating

to St. Joseph special graces, in order to render him apt for his sublime vocation. Let us congratulate this great saint on the precious gifts which he received from God and on his fidelity in corresponding with them. Let us honor him as the friend of God and privileged above all men who have ever been or ever will be.

FIRST POINT.

The Love of Jesus and Mary, the First Grace attached to the Devotion to St. Joseph.

If, according to the pious belief of the Church, the saints in heaven have a special power of obtaining for those who invoke them the graces in which they have most excelled, what power does not St. Joseph possess of obtaining the love of Jesus and Mary? For how greatly he loved both the one and the other! Alas, without St. Joseph, what would have become of the Child Jesus? Not having a father in this world, and forsaken by His true Father who was in heaven, He was about to become an orphan amongst men. Without Joseph, what would have become of Mary? A butt to calumny, without a consoler in her afflictions, without a guide in her travels, without a helper to share with her in the charge of providing for the subsistence of the Child, she was about to experience all the anguish of unhappiness. But, O Joseph! thy love took care of

Jesus and Mary; thou wert their guardian, their tutelary angel. When Mary carried in her womb the Incarnate God, thou didst protect the living tabernacle filled with the glory of the Most High; when she had troubles, thou didst console her; when she had needs, thou didst solace them; and when the Word made flesh appeared upon earth, thou didst become by affection that which thou wert not by nature; and thou didst lavish on Him the tenderness and the care of the best of fathers. On the day of the Circumcision thou didst give Him the name of Jesus; on the day of the Presentation thou didst take Him to the temple and didst place Him on the altar; in the days of His infancy thou didst receive His sighs, dry His tears, lovingly caress Him; thou didst shelter Him from the rage of Herod, thou didst feed Him at the cost of thy labors, and thou didst provide for all His wants. O most blessed Joseph, how thou didst love Jesus and Mary! How powerful must thou be to obtain for us this double love! Beg earnestly that we may have it, we conjure thee with all the fervor of our soul.

SECOND POINT.

The Gift of Purity and of Innocence, the Second Grace attached to the Devotion to St. Joseph.

What in fact can we conceive purer than St. Joseph? Would Mary have ever consented to live with a man, to associate her life with his, to exchange with him her words and her feelings, to allow him to carry in his arms the thrice-holy God, if that man had not been an angel of purity? Oh, how chaste and modest he must have been in his thoughts and affections, in his looks and in his senses, to be able to be admitted to live in the company of her who was purer than all the angels and all the saints put together, to be constituted the guardian of her purity and to partake with her in the care to be taken of the Infant-God. These are lofty mysteries which surpass the intelligence of even the seraphim; whence we must conclude that St. Joseph is the true patron of purity, and that whoever desires to lead in a body of flesh a life which has nothing in common with the flesh, ought to place himself under his protection. Are we faithful in recommending to St. Joseph our purity and our innocence, in invoking him in temptations, in proposing him to ourselves as our model in guarding this holy virtue, in keeping our senses, our mind, and our imagination in restraint?

THIRD POINT.

The Interior Life, the Third Grace attached to the Devotion to St. Joseph.

What saint, in fact, has excelled St. Joseph in the interior life? His life had nothing outwardly brilliant in it ; all its beauty, all its glory was within (Ps. xliv. 14). Dead to the world and its vanities, to its novelties and its pleasures, he occupies himself solely in adoring, contemplating, and enjoying God in the secret of his heart. To love Jesus and Mary, to converse with Jesus and Mary, that was his whole happiness, the whole of his glory, and, apart from that, the whole world is nothing to him. Oh, how powerful St. Joseph must be to obtain for his clients the grace of the interior life! Do we often ask it of him? Do we keep ourselves on our guard against dissipation of thought, and do we aspire after becoming really interior?

FOURTH POINT.

The Gift of a Holy Death, the Fourth Grace attached to the Devotion to St. Joseph.

St. Joseph died in the arms of Jesus and Mary; how ravishing must it have been thus to die! and how such a holy end must have enabled St. Joseph to obtain great power in gaining for his clients the grace of a holy death! How consoling a thought, and how well suited to inspire us with a tender devotion towards St. Joseph, for the grace of a good death is the grace of graces; it is the grace which decides whether we shall enjoy eternal happiness or be condemned to eternal woe, heaven or hell; it is, consequently, the grace which we should have it most at heart to obtain, that we ought to ask for every day, with more and more earnestness; and, if it were for this reason alone, it would be enough to attach our hearts to the devotion to St. Joseph, as patron of a good death.

Resolutions and spiritual nosegay as above.

March Twenty-second.—St. Joseph.

(FOURTH MEDITATION.)

Summary of the Morrow's Meditation.

It is not enough to honor and to invoke St. Joseph, as our preceding meditations have taught us to do: we must also imitate him; and that we may do so we will study the characteristics of his most holy life. The first characteristic, which will be the subject of our meditation to-morrow, is, that it was a hidden life,—a life hidden: 1st, in retreat; 2d, in silence; 3d, in obscurity and forgotten by the world. We will then make the resolution: 1st, not to go into the world, except from necessity; 2d, never to say anything which is to our own advantage; 3d, not to endeavor to parade ourselves and attract notice. Our spiritual nosegay shall be the words of the Imitation: *"Love to be ignored and counted as nothing"* (L Imit. ii. 13).

St. Joseph.

Meditation for the Morning.

Let us adore the Holy Spirit depicting to us a feature in the life of St. Joseph. He was a just man, the gospel says (Matt. i. 19), that is to say, perfect, and so remarkable for holiness that several holy doctors believe that he was, like St. John Baptist, sanctified in the womb of his mother. Let us ask this great saint for some share in his grace, above all in his humility, the base of all virtue, which made him so much love a hidden life.

FIRST POINT.

St. Joseph Leads a Hidden Life in Retreat.

This admirable saint does not go into the world excepting when he is obliged to do so; he goes to Bethlehem when the edict of the emperor constrains him; into Egypt when a heavenly command calls him thither; to Jerusalem when a religious duty invites him there. Beyond that, he appears nowhere. He is never seen in the town mingling in conversation and in worldly joys, in society and at the feasts of the children of men; his delight is to be in his beloved solitude of Nazareth. It is there that he enjoys God and His Divine Son, whilst occupying himself with the duties of his state; it is there that he spends his days sweetly, recollected in God and occupied with His sanctification. Let us learn from his example not to love the world, which dissipates and seduces the heart; to cherish a life of retreat, in which we study and know ourselves, where solid virtues are formed, and where we become accustomed to that calm, to that interior recollection, without which all progress in piety is impossible. Let us recall to ourselves the words of the wise pagan who said: "*Every time that I have gone amongst men, I have come back less a man than I was*" (I. Imit. xx. 2), and the words of St. Leo the Great: "*The dust of the world necessarily soils even the most religious hearts.*"

SECOND POINT.

St. Joseph Leads a Hidden Life in Silence.

St. Joseph was descended in a direct line from the greatest kings of Juda, and from the most illustrious patriarchs. The guardian of the secrets of the Most High, he lodged in his house a God, who honored him with the title of father. Nevertheless, humility made him keep silence respecting so lofty a birth, so much greatness, and so much glory. Others would have imagined that it would be necessary, at any rate for the honor of Jesus, to divulge the nobility of his origin, to make himself the apostle and the evangelist of the Divine Child, in order that all might come and adore Him; but Joseph, wiser in humility, felt that it was better to be silent than to say a single word which would be to his own advantage, and left it to God to make His Son to be known. Not a single neighbor, not a friend is admitted into the secret; and at the end of thirty years, the Son of the Eternal Father is only known as an artisan, and the Son of the carpenter Joseph (Matt. xiii. 55; Mark vi. 3). O marvellous silence! Joseph has in his house what was sufficient to attract the eyes of the whole world; and the world knows nothing of it; he possesses a God-Man, and he does not say a word about it. The Magi and the shepherds come to adore Jesus; Simeon and Anna proclaim His greatness, and Joseph says nothing to them: Joseph who had been so thoroughly instructed by the angel with regard to the divinity of the Child, Joseph who knew, because he had been a witness to it, the miracle of his birth. What father would not have spoken of such a son? Joseph faithfully keeps his secret, and carries it to his grave. A beautiful lesson, which teaches us never to say, or even to hint anything to our own advantage, and never to take vanity as our counsellor in whatever we may say.

THIRD POINT.

St. Joseph Leads a Hidden Life in Obscurity and Forgotten.

This man, so great before God, is unknown to

the whole world; it is much if a few neighbors know him by the title of a poor artisan. He who was made to occupy the throne of kings inhabits a poor dwelling; he who might have asked for honor and glory from Him who disposes at His pleasure of sceptres and empires, prefers to remain in obscurity. If we would know the reason of this preference we have only to ask him, and in the bottom of our hearts we shall hear him reply: "Every day I saw before my eyes, subject to the weakness and abasement of infancy, Him whom all heaven adores with trembling; Him who has but to say a word and the angels, swifter than lightning, go to execute His commands; and seeing all this, the love of glory and of celebrity, the desire to parade one's self and to attract notice seems to me to be folly; a hidden life, in which one is forgotten appears to me to be the only real glory, and I bless Heaven to have made me descend from the lofty rank occupied by my ancestors; I prefer my humble dwelling to the palaces of kings, my food and my coarse raiment to the purple of the great and the delicate viands on their tables." We may, indeed, easily conceive that a man of the world, who has not meditated upon Jesus Christ, and who sees nothing beyond the tomb, runs after the rumor which is called renown, and endeavors to aggrandize his littleness by exterior pomp;

but the Christian who has reflected for a few moments upon the abasement of the Incarnate Word, and who beholds, beyond this life, a new world and a new heaven, where, in exchange for being forgotten and being obscure here below, he will have glory which is at once incomparable and immortal, such a Christian ought to disdain, as being quite below him, so false a possession, so temporary as is the glory of the world and the esteem of men, and aspire only to the glory of eternity. Are these our dispositions?

Resolutions and spiritual nosegay as above.

March Twenty-third.—St. Joseph.

(FIFTH MEDITATION.)

Summary of the Morrow's Meditation.

We will meditate to-morrow upon the second characteristic of the life of St. Joseph; it was a tried life, and tried just by those who were dearest to him; his charity was put to the proof by Mary, his faith by Jesus, his obedience by God the Father, his patience by Providence. After these four considerations we will make the resolution: 1st, to receive all the trials of life with calmness and peace, without allowing ourselves to give way to impatience, to murmurs and discontent, and lovingly to obey in all things the good pleasure of God ; 2d, carefully to avoid rash judgments with regard to our neighbor, and want of faith with regard to God. We will retain as our spiritual nosegay the words of the Apostle : *" The sufferings of this time are not worthy to be compared with the glory to come"* (Rom. viii. 18).

Meditation for the Morning.

Let us adore the great design of God, who exposes His dearest friends to trials (Job xii. 13). We find a difficulty in understanding, here below, this arrangement of Providence (Acts xiv. 2), but let us have patience, and we shall understand it hereafter (John xiii. 7). Meanwhile let us adore without understanding, and let us lovingly bless God, who does all things well (Mark vii. 37).

FIRST POINT.

The Charity of St. Joseph put to the Proof by Mary.

Mary, after the Incarnation of the Word, could not either hide her situation from St. Joseph, or explain to him the mystery of it, since the slightest enlightenment would have turned to her praise. What a trial for the charity of Joseph. On one side he will not either condemn or suspect

Mary: charity thinks no evil and judges no one; on the other side, he asks himself what this mystery can be. How is it that Mary does not herself say something to tranquillize so cruel an anxiety? In his embarrassment, he was about to separate himself from her in accordance with the precepts of the law, when an angel descended from heaven to reveal to him this mystery, and to change the trouble of his soul into a delightful sentiment of admiration, of respect, and of love. Is it thus that we observe reserve in our judgments with regard to our neighbor? Do we not often permit ourselves to indulge in evil suspicions respecting the intentions and the designs of our brethren? and, what is worse still, do we not communicate to others our bad impressions? Do we take care in as far as we can to cast a veil over the faults of our neighbor, to excuse them, to hide them, and to turn the conversation when others speak of them?

SECOND POINT.

The Faith of St. Joseph put to the Proof by the Word Incarnate.

In the state in which Jesus showed Himself to Joseph, what was there which announced Him to be a God? Like other children, He was feeble and delicate, His limbs were powerless and weak, from His eyes tears escaped, and wailings from His mouth. How amidst so much weakness was it possible to recognize the God who by one word gave being to everything that exists ; who by one single glance makes the world tremble ; who with His hand measures the breadth of the heavens and with a finger sounds the depth of the seas ? Again, during the three last years of His life He sustained His seeming weakness by miracles ; but here nothing gives any sign of His power, nothing but weakness alone, forsaken, even unnatural, if we are to judge by appearances, for it was necessary to shelter this Child by flight from being pursued by a man ; it was necessary to save Him under cover of the darkness, and He could not return until those were dead who had sought the life of the Child, even as though a God would not be in safety as long as they were alive. What a trial for the faith of Joseph ! But he comes out of it a conqueror, and without asking for miracles like the incredulous Jew, he recognizes his God in this Child ; in this external powerlessness he adores the supreme power which commands everything. The annihilations of the Incarnate Word increase his charity, which admires the principle of them (St. Bernard), but without in any degree diminishing the vivacity and simplicity of his faith. What a lesson for us in presence of the Holy Eucharist !

THIRD POINT.

The Obedience of St. Joseph put to the Proof by God the Father.

Go to Bethlehem, He had said to him, by means of the command of a heathen king; fly into Egypt, an angel came to tell him; return to the land of Israel, another heavenly messenger came to say to him. But, O my God, what could not Joseph have brought forward in opposition to all these commands? To go to Bethlehem! but Mary was approaching the term of her pregnancy, and the journey would expose to peril the life of the Child. To fly into Egypt! but what can a God fear from a man? To fly that same night! but why not wait for the daylight? To fly to so distant a country! but with what resources sufficient to defray the expenses of the journey and the needs of three persons in an unknown land? Joseph does not hesitate in presence of these trials; he obeys with simplicity, he sets off instantly and without questioning. O prompt and perfect obedience, simple and upright, courageous and intrepid! What a fresh lesson for me!

FOURTH POINT.

The Patience of St. Joseph put to the Proof by Providence.

We do not read that before the birth of Jesus Joseph was exercised by great tribulations; he doubtless led the kind of humble and modest life which finds its happiness in what suffices. But after the birth of the Divine Child, the life of Joseph was nothing but one long martyrdom; until then he had not been without a home; afterwards, his retreat was a stable. Until then, he had lived tranquilly, surmounting his poverty by labor; afterwards, he was persecuted, and constrained to lead a life of exile in a strange land. Until then, desiring little, he had known but little anguish; afterwards, his compassionate soul was torn when he heard the old man Simeon say to Mary: "*Thy soul a sword shall pierce*" (Luke ii. 35); until then, possessing little, he had little to lose; afterwards, he had Jesus, and he lost Him at Jerusalem. What an affliction! Ah, rather have lost all, for without Jesus, what is the whole earth? He does indeed find Him at last, but death comes to separate him from Jesus a second time, and he quits Him, not to see Him again until, on the day of His resurrection, He visits limbo. It is thus that the patience of St. Joseph was tried by a series of tribulations, and, in the midst of it all, Joseph was calm and resigned. He understood that tribulations are the crucible in which God purifies the virtue of those whom He loves, that the way of the cross is the only one which leads to heaven, that all the just must pass along it, and that Jesus never visits a soul without taking His cross with Him. Are these our dispositions? Do we not imitate, on the contrary, the man of the world who lives only for pleasure, who will always have his comforts and be at his ease, who will deprive himself of nothing, mortify himself in nothing, and who is irritated by suffering and contradiction?

Resolutions and spiritual nosegay as above.

March Twenty-fifth.—The Feast of the Annunciation.

The Gospel according to St. Luke, i. 26-38.

"And in the sixth month the angel Gabriel was sent from God into a city of Galilee, called Nazareth, to a virgin espoused to a man whose name was Joseph, of the house of David; and the virgin's name was Mary. And the angel being come in, said unto her: Hail, full of grace, the Lord is with thee: blessed art thou among women. Who having heard, was troubled at the saying, and thought with herself what manner of salutation this should be. And the angel said to her: Fear not, Mary, for thou hast found grace with God. Behold thou shalt conceive in thy womb, and shalt bring forth a Son: and thou shalt call His name Jesus. He shall be great, and shall be called the Son of the Most High, and the Lord God shall give unto Him the throne of

David His father: and He shall reign in the house of Jacob forever, and of His kingdom there shall be no end. And Mary said to the angel: How shall this be done, because I know not man? And the angel answering, said to her: The Holy Ghost shall come upon thee, and the power of the Most High shall overshadow thee. And therefore also the Holy which shall be born of thee shall be called the Son of God. And behold thy cousin Elizabeth, she also hath conceived a son in her old age; and this is the sixth month with her that is called barren; because no word shall be impossible with God. And Mary said: Behold the handmaid of the Lord, be it done to me according to thy word. And the angel departed from her."

Summary of the Morrow's Meditation.

We will consider, in the mystery of the Incarnation: 1st, a mystery of love; 2d, a mystery of humility. We will then make the resolution: 1st, to renew our love of the Word Incarnate, and of the devotion to the *angelus*, which is the pious memorial of it; 2d, to practise humility, whether in our language, by keeping silence respecting all which might tend to our praise, and suffering in silence all that wounds us; whether it be as regards our manners, our clothes, or the whole of our exterior. Our spiritual nosegay shall be the words which St. John brought down to us from heaven: "*The Word was made flesh*" (John i. 14).

Meditation for the Morning.

Let us transport ourselves in spirit to the venerable oratory where Mary, at prayer, received the visit of the archangel Gabriel. To the enunciation of the heavenly will Mary replies by a humble compliance: "*Be it done to me according to thy word.*" Oh, how efficacious is this *fiat!* It is a word of submission and obedience, but it is more powerful than the word of command by which the world was created; for it gave being to the Creator Himself, and reconciled heaven with earth. Hardly had she uttered it than by the operation of the Holy Ghost the Eternal Word was incarnated in the womb of Mary, and Mary became the Mother of God. Let us prostrate ourselves in presence of these lofty mysteries; let us adore, let us admire, let us love.

FIRST POINT.

The Mystery of the Incarnation is a Mystery of Love.

In order to penetrate ourselves as we ought with this truth, let us first consider who He is who comes from heaven to save us. God does not send one of His angels to succor us; He comes Himself in person; Himself, the true Son of God, the true God, the Life Eternal, the beginning and the end of all things; Himself, the splendor of the Father, the image of His substance, in whom inhabits the plenitude of the Divinity, and who does no injury to God, His Father, in saying that He is equal and consubstantial to Him. He comes, and how? By taking all that is vilest in us, and that merits only hatred and confusion; by making Himself flesh and clay like us; that is to say, the Omnipotent makes Himself weak, the Eternal makes Himself mortal, the Ancient of days takes a commencement, the Immense renders itself little, a God makes Himself man. Creation is doubtless a great work of love; the preservation of our being, which is, as it were, a creation of every moment, is not less so; but how much more is love revealed in the Incarnation! In order to create the world, it only cost God a single word; in order to preserve it, an act of His will; but here He descends from the heights of His eternal throne down to the lowest stage of this inferior world, there where He finds the most humiliation and the greatest suffering. O incomprehensible love! He comes thus; and for whom? For man, that being so little, that creature so weak, who crawls in this lower world at an infinite distance from His throne; for man, fallen from the dignity of his nature, poor, naked, despoiled, so that he is ashamed of himself (Gen. iii. 10); for man, who had made himself His enemy; for this worthless rebel, him whom justice demanded to strike, when mercy cast itself before him in order to ward off the blows; for man, whose insensibility, ingratitude, and relapses He foresaw, together with the malice which made him tread under foot even the blood of his God, the obstinacy, carried to excess, which makes him lose his soul, spite of all the means provided for his salvation. He comes; and at what a time? When this whole world, disfigured by hideous crimes, corruptions, and errors, was rolling itself in the mire of all the vices, and was adoring deified passions, God the Father loved this hideous world so incomparably as to give it His only Son (John iii. 16). God the Son loved it so incomparably as to become flesh for it. He might have left us all under the anathema of eternal damnation, like the rebel angels; but, O ineffable love! when we were His enemies, and when He owed us nothing but vengeance, He came to us from pure goodness and wholly gratuitous mercy (Eph. ii. 4). And what did He bring with Him? He brought us all blessings and graces, for Jesus Christ is all such (Coloss. iii. 14). By the Incarnation human nature is raised above the angels, since by it we may say that man is God and God is man. By the Incarnation we pass from the slavery of the devil to the rank of children of God, brothers of a God, heirs of heaven, co-heirs with Jesus Christ of an eternal kingdom. By the Incarnation our sins are pardoned, a single act of contrition introduces us into heaven, all our prayers and our good works acquire infinite value. Lastly, by the Incarnation we have light to know the truth, the highest examples to lead us to what is good, grace to enable us to act, strength to sustain us. Are not these abysses of love? And what soul, even though it were made of iron, would not be touched? (St. Bernard.)

SECOND POINT.

The Mystery of the Incarnation is a Mystery of Humility.

First of all, what humility in Mary! The angel calls her "*full of grace*"; and she esteems herself nothing more than a poor, indigent person, who possesses nothing but what she has received from the Lord. The angel calls her "*blessed among women*"; and she looks upon herself as a woman of no account, whom God has raised through pure goodness. 2d. The angel says to her, "*Thou hast found grace before God*"; and she replies: It is because He has had regard to my lowliness. Lastly, the angel tells her: "*Thou art the mother of God*"; and she replies: I am His servant. Such great humility made her become at that very moment, says St. Bernard, the mother of God. Oh, how true it is, then, that the waters of grace descend into humble souls like the rains of heaven into deep valleys, and that as precious metals are found hidden in the bowels of the earth, and pearls in the depths of the sea, so it is that in humble souls God founds the loftiest virtues. Humility so greatly pleases God that, in coming upon earth, He made it His own special virtue. In order to understand it, let us rise above the highest heavens to that sublime solitude where the infinite excellence of His Being places Him— at an incommensurable distance from all created beings. This will be the starting-point which will enable us to measure the humility of the Incarnate Word. He descends first to the dazzling order of the seraphim, which, for a God, is already an immense descent; it is to traverse the infinite; He still descends, and descends until He arrives at our nature. It is in our clay He wills His majesty to be. But in this clay there are different degrees. There is the clay which shines beneath the splendor of gold and of purple. It is doubtless a false splendor, but yet it shines; the Word of God will have none of it. He, therefore, descends yet lower; first, He finds a stable, then the dwelling of an artisan; He finds a poor woman, who gains her bread by labor. He descends even lower than this, and He hides Himself in her womb; He chooses this obscure person to be His first dwelling upon earth. O abyss of humility! Who, after this, would desire esteem and glory? who would wish to appear in public, to attract notice, to make himself applauded? Who would not love a hidden life?

Resolutions and spiritual nosegay as above.

See the meditations on the saints at the end of the third volume for meditations on St. Mark (April 25th), the month of Mary (April 30th), SS. Philip and James (May 1st), and the feast of the Invention of the True Cross (May 3d).

www.ingramcontent.com/pod-product-compliance
Lightning Source LLC
Chambersburg PA
CBHW050845300426
44111CB00010B/1130